Table of Contents

History of CNC Objectives

1. The student will know the concept of CNC machining.

2. The student will know the year NC was invented, and by whom.

3. The student will know the benefits of CNC technology.

LearnCNC ™

for Haas Machine Tools

Lathe CNC Programming

An introduction to codes and programming, this manual is designed for beginner to intermediate level *lathe* CNC operators and programmers. The content and sample programs provided cover a broad range of CNC programming requirements.

Basic mathematics and formulas are used.

Learning materials may be used in conjunction with optional Virtual Training Environment for CNC including virtual CNC

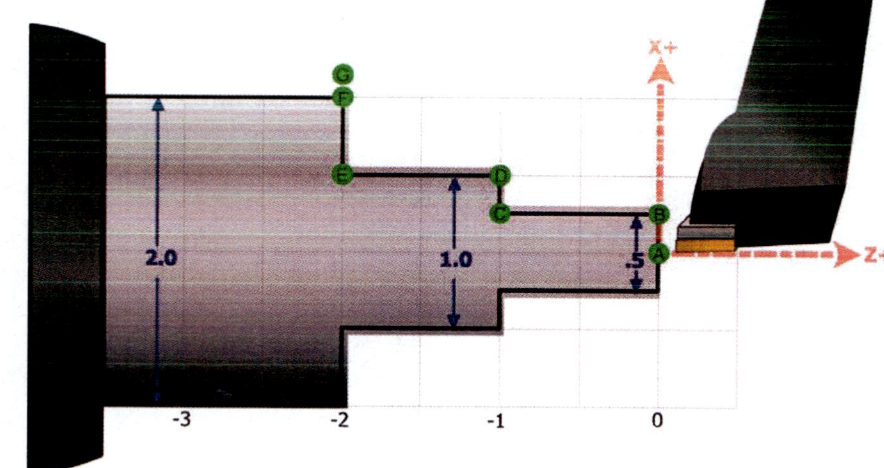


```
  G70 P10 Q20
  N10 G42 G0 X0 Z.2
A G1 F.005 Z0
B X0.5 Z0
C X0.5 Z-1.0
D X1.0 Z-1.0
E X1.0 Z-2.0
F X2.0 Z-2.0
G N20 G40 X2.1 (linear motion to clear X)
```

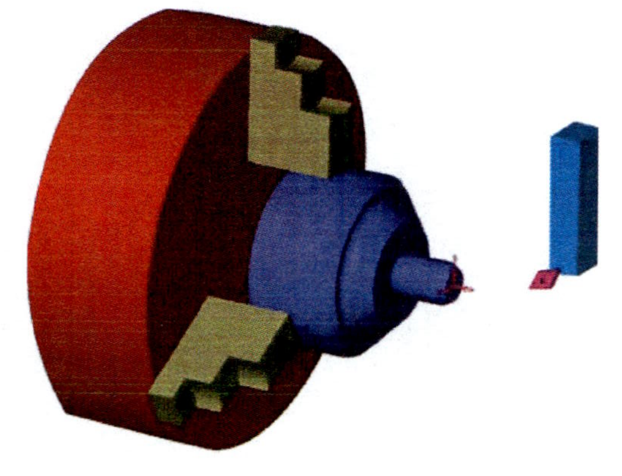

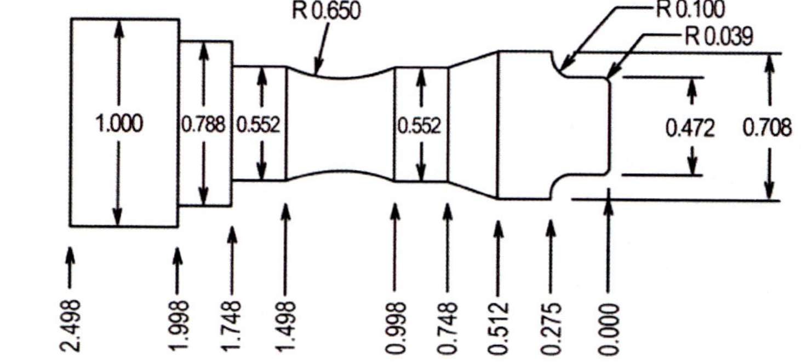

ISBN-13: 978-0-9816982-2-9

LEVEL 1

Powered by immerse2learn.com

Learning Productivity Tools

Virtual Training Environment
CNC Machining

Flight Simulator Technology

The Virtual Training Environment for CNC Machining combines *powerful* "flight-simulator" technology with a *flexible* Internet-based learning content management system to deliver a truly innovative learning experience.

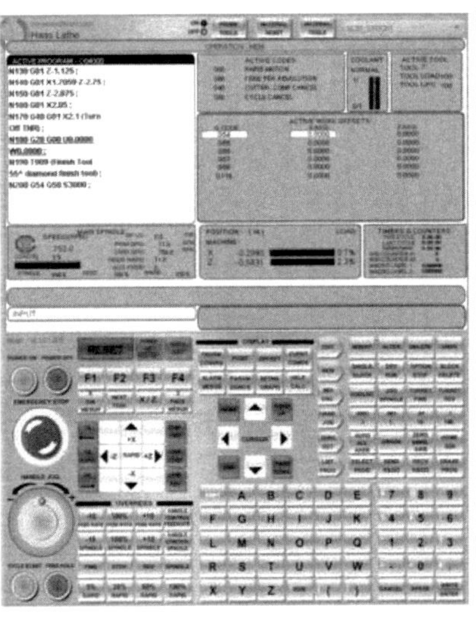
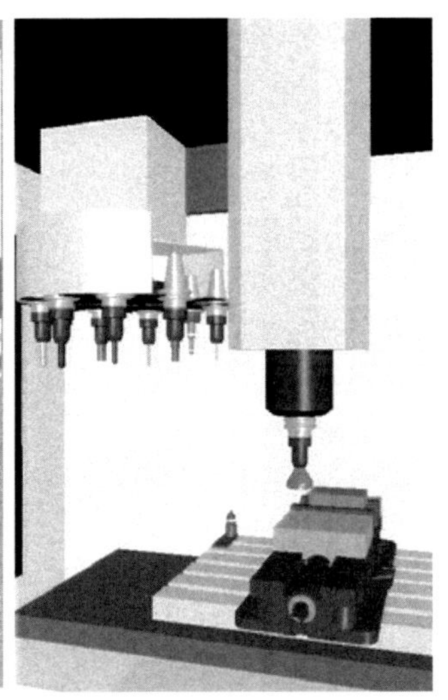

Simulated 3D machine models

Mimic Reality

- Virtual mills and lathes
- Industrial control panels
- Edit, load, run and save NC programs
- Set tool and work offsets
- Touch probes
- Stock material removal
- Canned cycles
- Control alarms

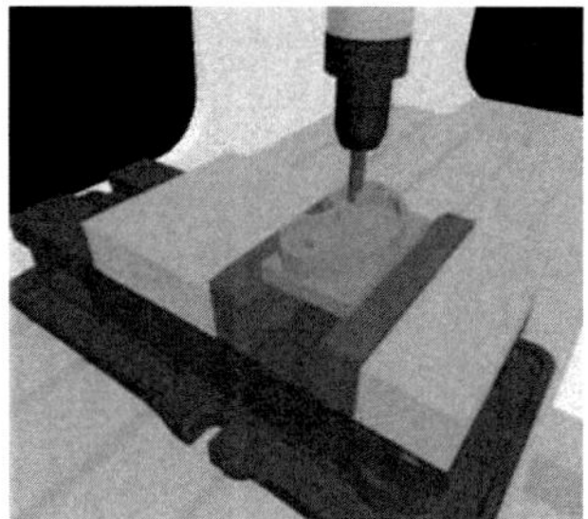

Real-time material removal

Mill and lathe touch probes

Learning Productivity Tool

Unlimited access to train and rehearse in the Virtual Training Environment for CNC machining enables learners to develop greater confidence and proficiency prior to performing actual procedures and operating equipment.

Learn at Your Convenience!

Learn CNC machine setup and programming with popular control panels for Haas Machine Tools.

Major Benefits

- Cost-effective and safe
- 24/7 access at your convenience
- Track and measure learner progress
- Reduce risk to people and equipment
- Increase control panel training contact-time
- Learn with virtual controls and 3D machines

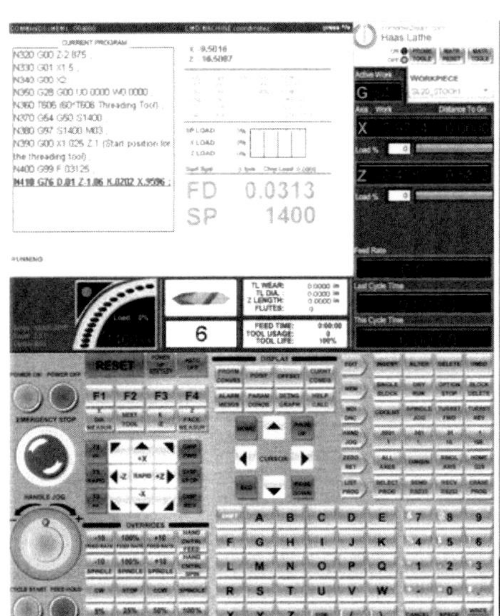

Popular control panels

Lathe applications

248.865.2001 www.immerse2learn.com Learning Productivity Tools

Additional Products

Online Learning Courses and Virtual 3D CNC Machine

Mill and *Lathe* CNC online courses provide the learner with comprehensive learning content, interactive exercises and virtual CNC panels and 3D machines. Depending on the course selected learning modules may include:

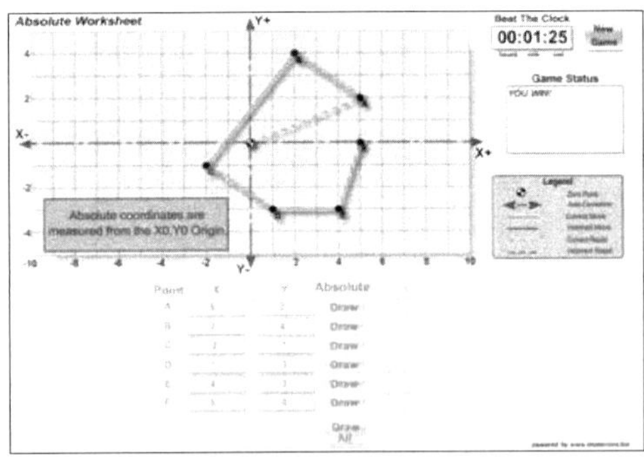

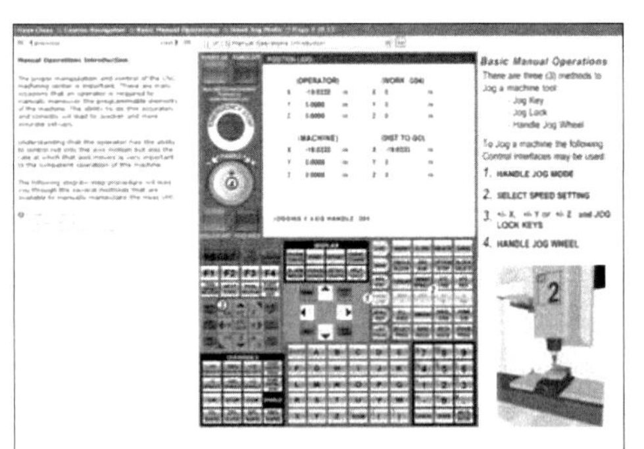

Setup

- Machine Motion Description
- CNC Panel Interface
- Machine Start-up
- Manual Operations
- Job Setup
- Edit Capabilities
- Program Entry
- Program Run

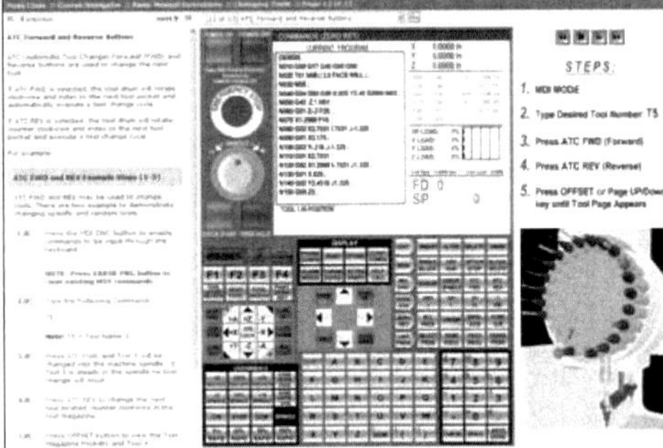

Programming

- Codes and Programs
- Program Structure
- Cartesian Coordinates System
- Cutter Compensation
- Tool Nose Radius Compensation
- Circular Interpolation
- Hole Manufacturing
- Programming Labs

Interactive online courses

Color Manuals, Exercises and Projects

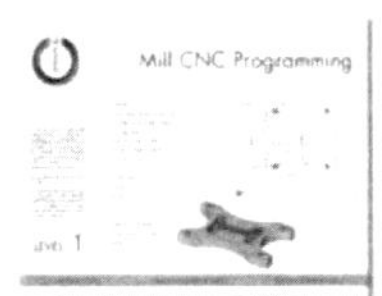

Mill CNC Programming Level 1

An introduction to codes and programming, this manual is designed for beginner to intermediate level *Mill* CNC operators and programmers. The content and sample programs provided cover a broad range of CNC programming requirements. Basic mathematics and formulas are used.

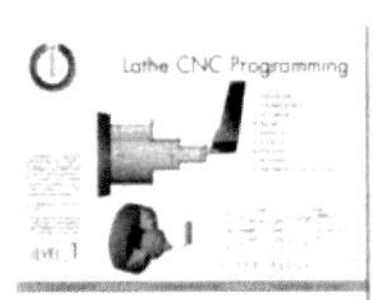

Lathe CNC Programming Level 1

An introduction to codes and programming, this manual is designed for beginner to intermediate level *Lathe* CNC operators and programmers. The content and sample programs provided cover a broad range of CNC programming requirements. Basic mathematics and formulas are used.

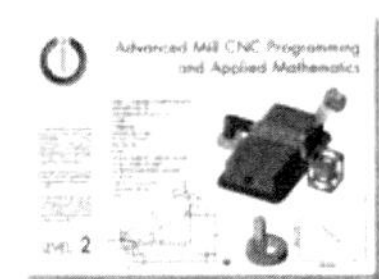

Mill CNC Programming and Applied Mathematics Level 2

This intermediate to expert level manual provides much more advanced application of codes, programming, and use of canned cycles. The content and sample programs provided cover a broad range of CNC programming requirements. Advanced mathematics and formulas including applied shop floor trigonometry and geometry are used.

Other Products and Services

- Robotic simulation
- Learning content management system
- Online evaluation tools
- Content development
- Onsite training
- CNC programming

History of CNC

FOCUS: This module focuses on manual programming of two dimensional parts for turning. It serves to build a foundation for CNC lathe programming.

HISTORY: In 1949 the Air Force asked John Parsons to develop a machine that would move each axis to specified point location automatically. John then asked Massachusetts Institute of Technology (MIT) to develop a motor that could control the axis of the motion. This became the servo. This combined technology became known as **Numerical Control.** NC technology was instrumental in the machining of airfoils for the Air Force. Later, with the invention of the computer and transistor circuitry, **Computer Numerical Control** (CNC) was born.

Although the software and hardware technology have drastically improved over the years, the basic principles today remain the same.

CNC technology offers many benefits over manual machining. Simple benefits include the ability to machine more than one part with consistency, and to contour circular features—something difficult to do manually. Complex benefits include High Speed Machining, Hard Milling and Turning, and multi-axis programming—all leading towards total automation.

Chapter 1
Introduction to CNC Turning

Objectives

1. The student will know the Cartesian Coordinate System as it applies to Turning.

2. The student will be introduced to circular and linear interpolation.

3. The student will be introduced to modal commands.

4. The student will understand Part Zero (ORIGIN).

Cartesian Coordinate System

By studying the part drawing and calculating dimensional values, a CNC program can be developed that maps the drawing coordinates onto the Turning Center for turning. Although only two axis of motion, lathe programming is different than mill programming due to the way coordinates are mapped and how cylindrical parts are machined.

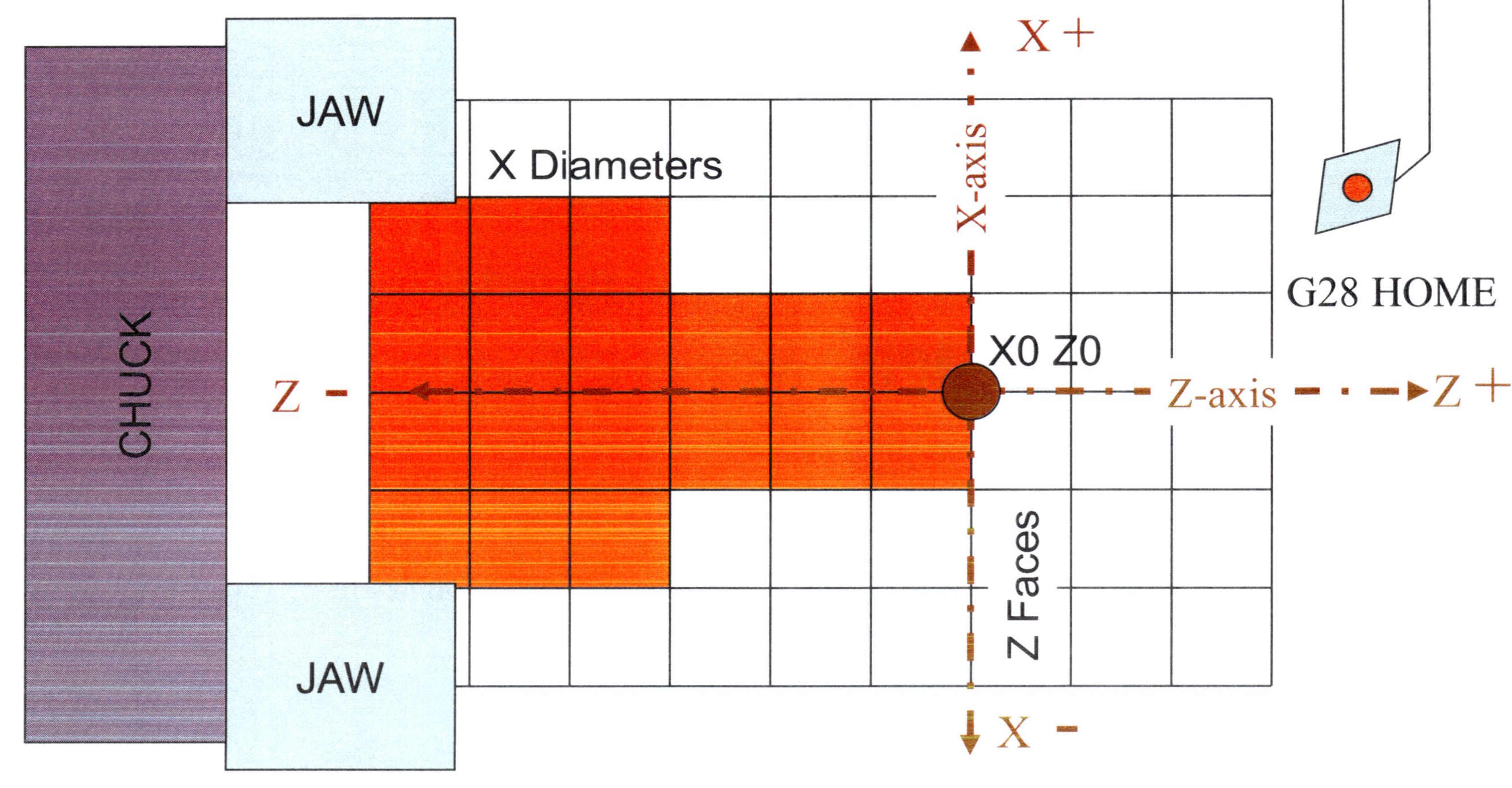

Linear and Circular Interpolation

CNC Turning is comprised of a series of linear and circular motion to rough and finish cylindrical parts. G0 and G1 are straight line motion from point to point, commonly referred to as **linear interpolation.** G0 is rapid motion. G1 is motion at feed rate.

G2 and G3 is circular motion, commonly referred to as **circular interpolation.**

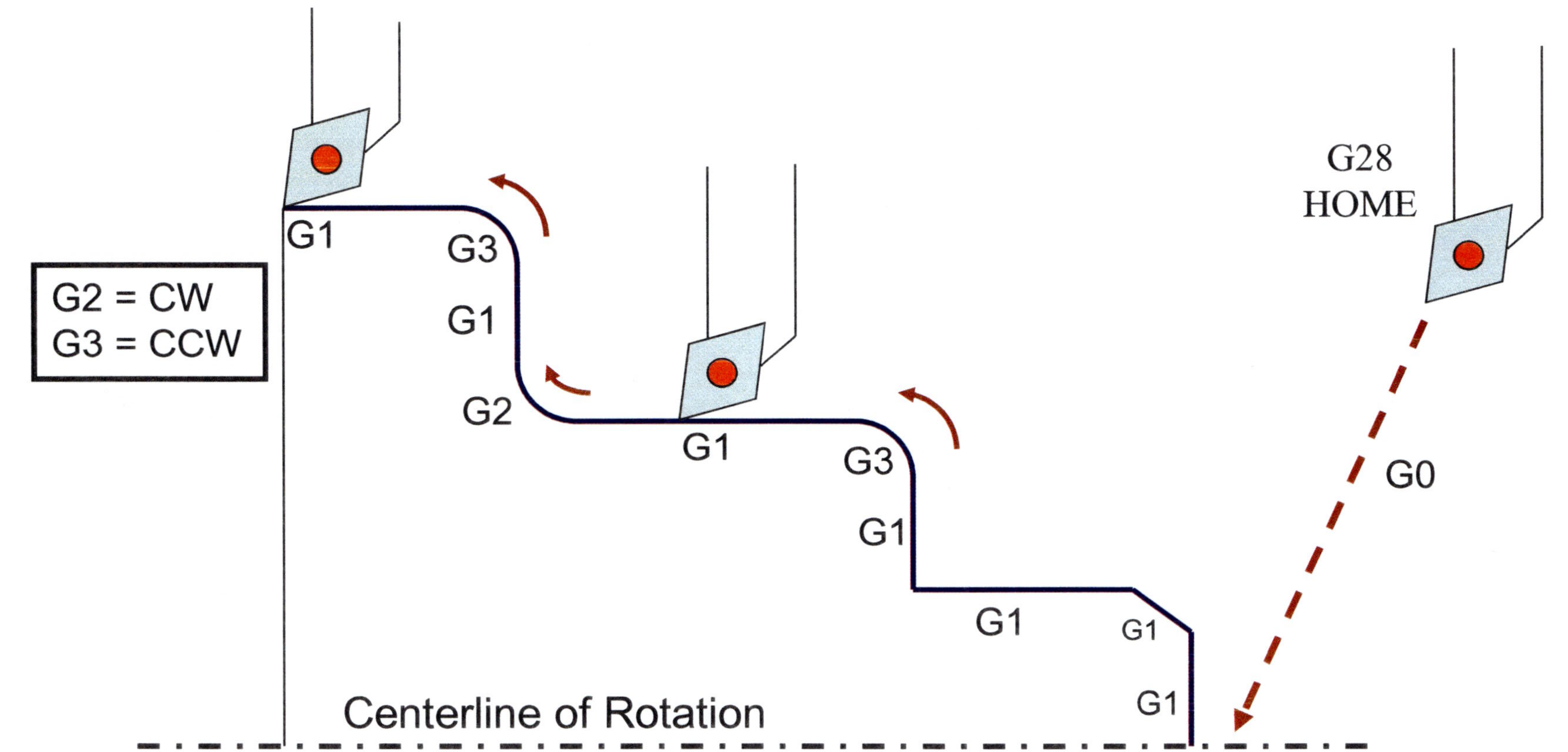

Modal and Non-Modal Commands

Modal commands are commands that remain active until made inactive by another command in the same group. **G0, G1, G2 and G3 are a group of modal commands.** When G1 is made active, it remains active until one of the other commands in the group is made active and turns off the G1 command.

Non-modal commands are active only in the block where they are used.

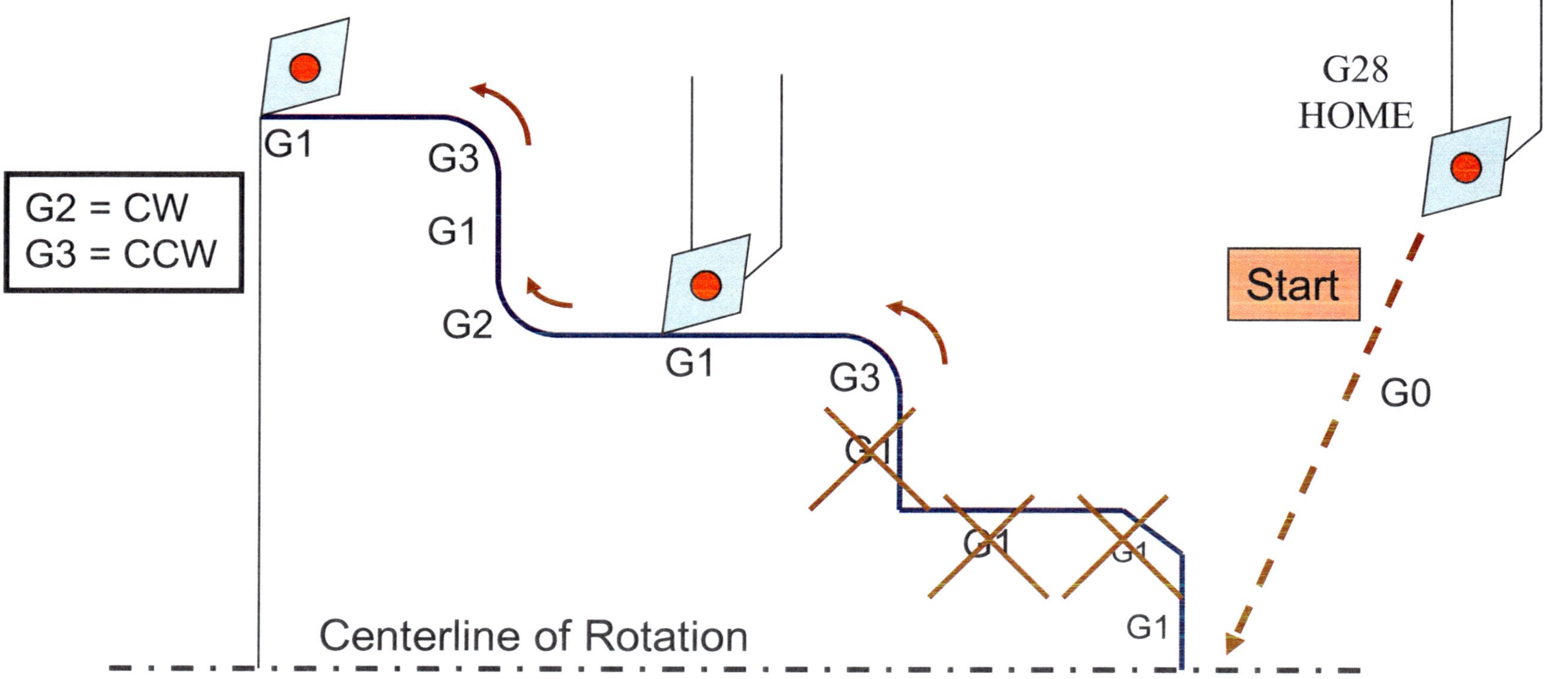

Part Zero

- Part Zero (Origin) defines X0 Z0 on the CNC lathe.
  - It can be located anywhere the programmer desires—usually dictated by the part print.
  - The machine tool operator locates this point on the machine and sets the values in the control panel.

- There is usually one part zero per setup.
  - The operator defines part zero by touching off with the tool diameter in X and the part face in Z.
  - These points are used to set the values in Work Offset page in the machine control.

Chapter 2
Mapping Print Coordinates

Objectives

1. The student will know how to map print coordinates to CNC Lathe coordinates.

2. The student will know how to transfer the coordinates to a CNC lathe program.

3. The student will be exposed to a complete CNC lathe program.

4. The student will engage in coordinate mapping exercises.

Mapping Print Coordinates

Coordinates are mapped (plotted) by looking at the imaginary grid underneath the part print.

Z values start at the Part Zero which is usually the face of the workpiece.

X values specify diameters. X0 is the centerline.

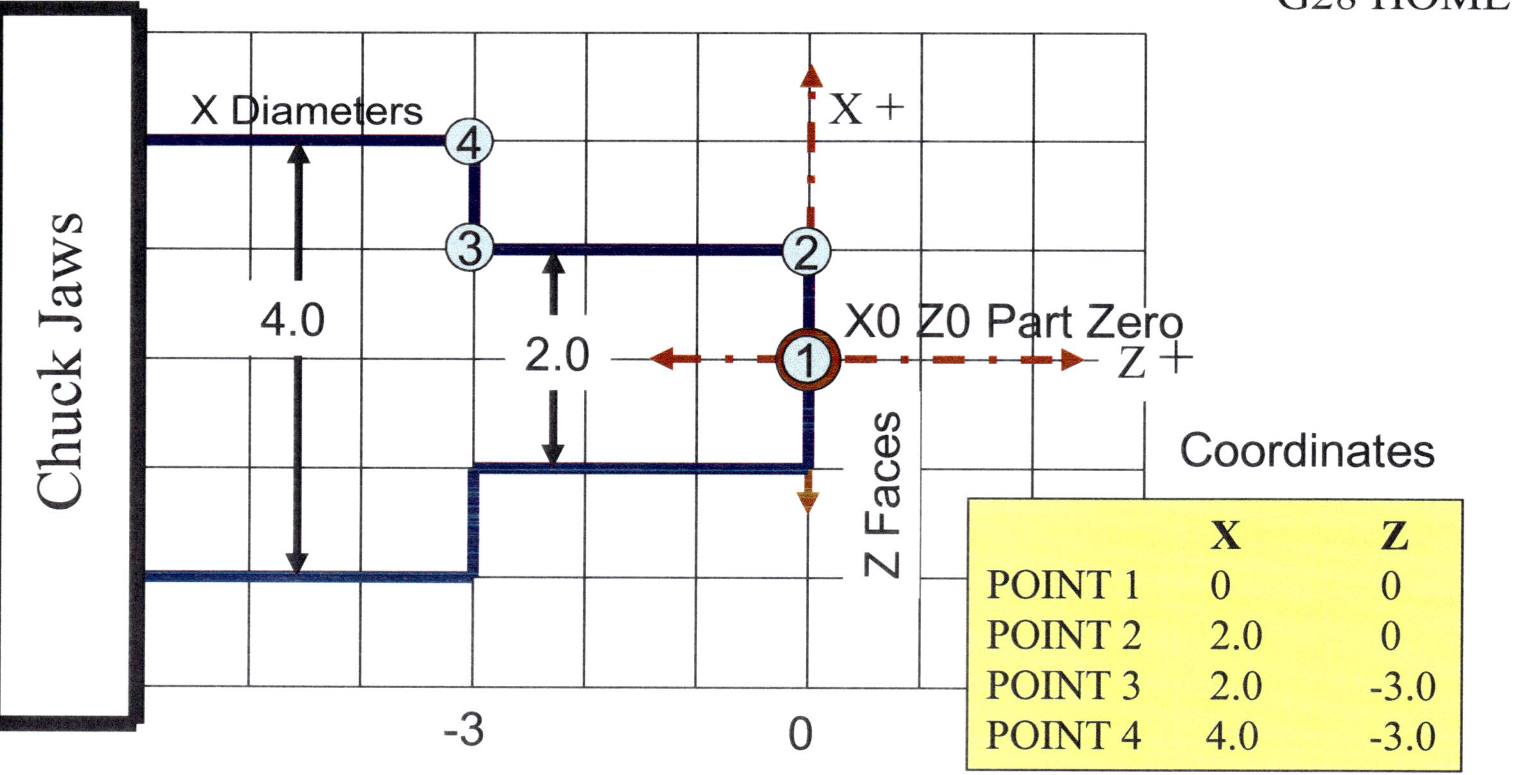

Coordinates

	X	Z
POINT 1	0	0
POINT 2	2.0	0
POINT 3	2.0	-3.0
POINT 4	4.0	-3.0

Mapping Worksheet 1

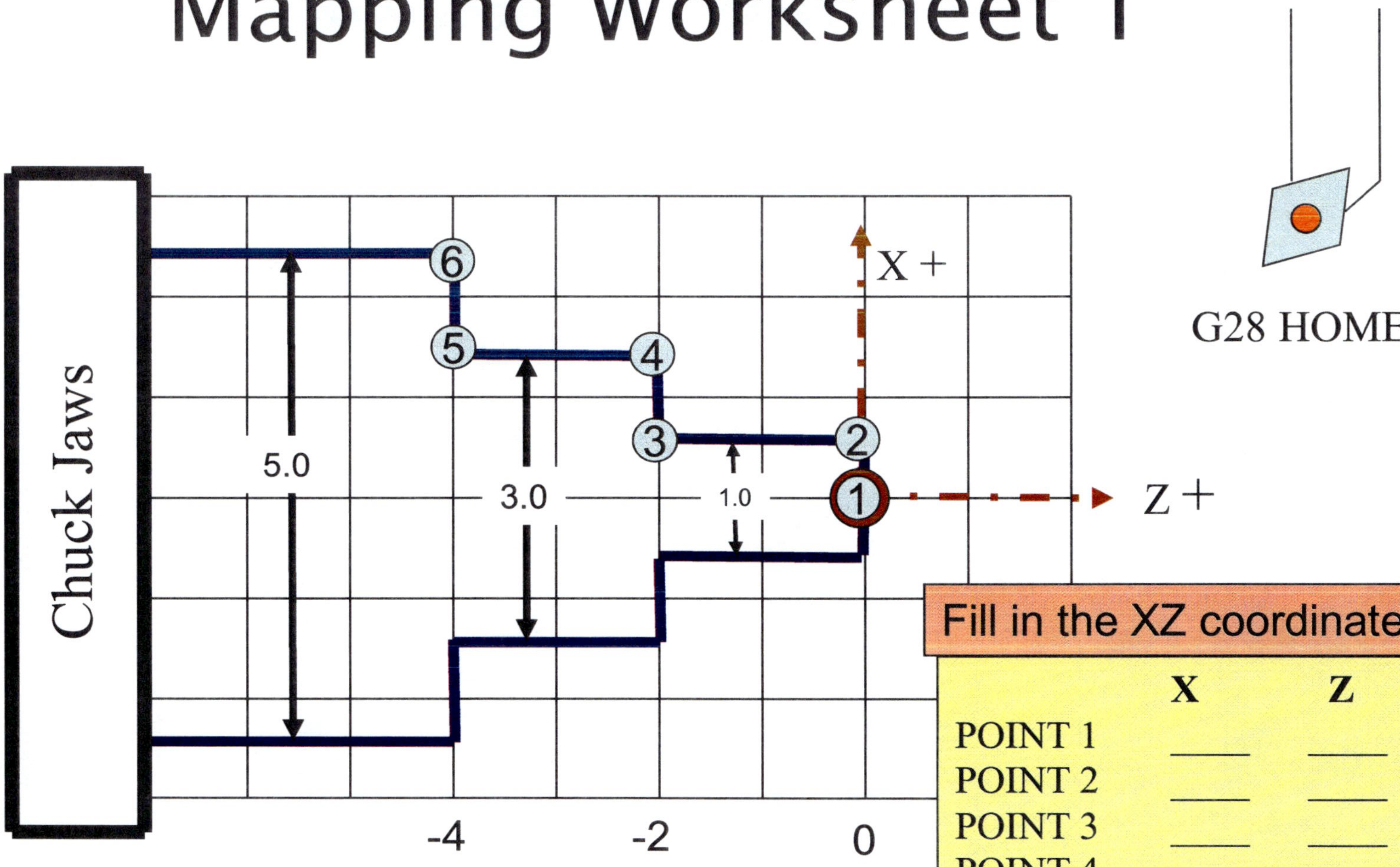

Fill in the XZ coordinates

	X	Z
POINT 1	____	____
POINT 2	____	____
POINT 3	____	____
POINT 4	____	____
POINT 5	____	____
POINT 6	____	____

Worksheet 1 Answers

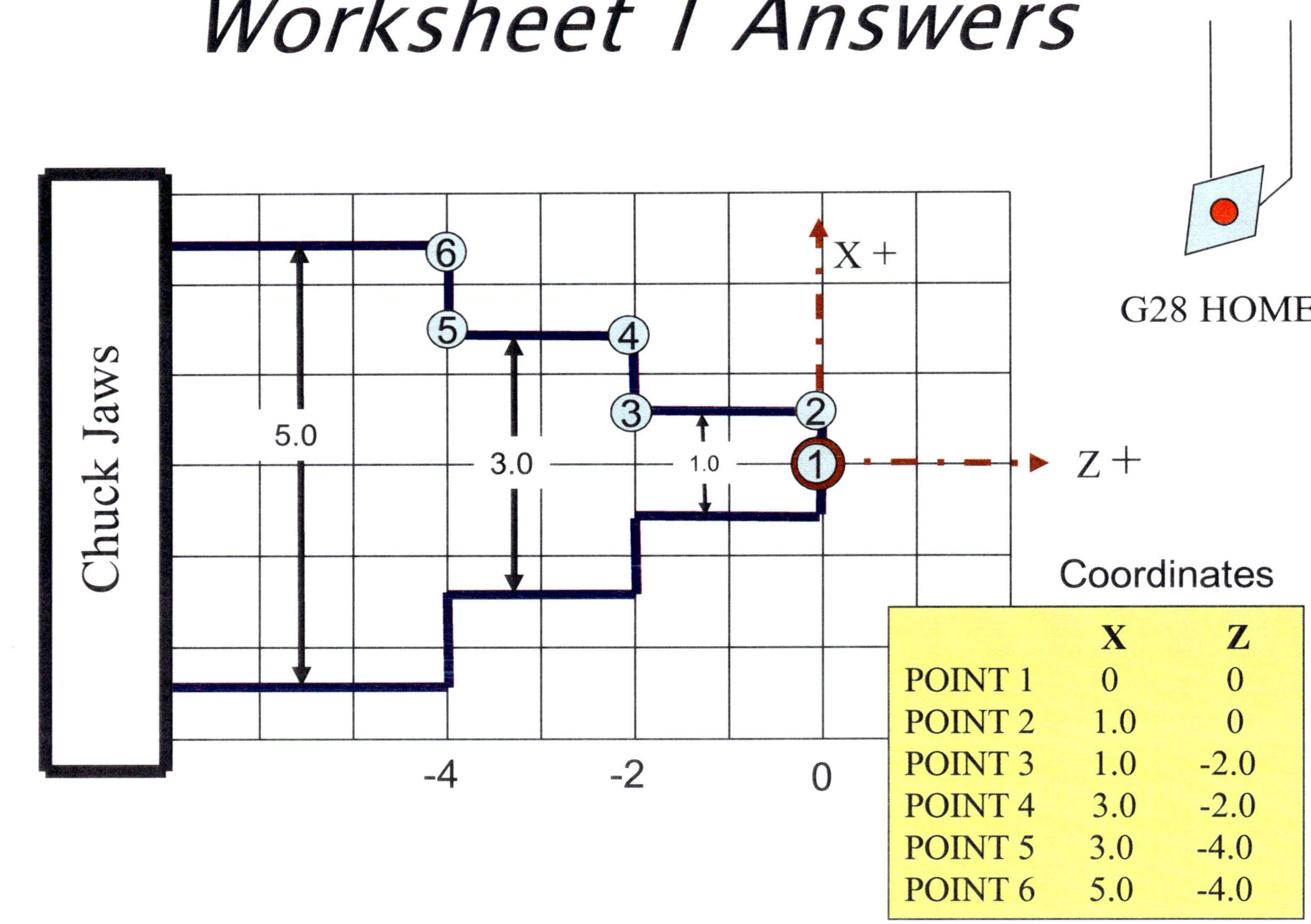

Coordinates

	X	Z
POINT 1	0	0
POINT 2	1.0	0
POINT 3	1.0	-2.0
POINT 4	3.0	-2.0
POINT 5	3.0	-4.0
POINT 6	5.0	-4.0

TURNING: LINEAR INTERPOLATION

- Linear Interpolation allows you to machine straight-line movements at programmed feed rates. It's commonly referred to as *point-to-point*.
  - The tool machines from the current position to the programmed position, commanded by a G1, a feed rate, and a desired location (X, and/or Z).
    - G1 X2.5 Z-0.125 F.01
    - G1 is **modal**. It stays in effect until cancelled by G0, G2, or G3.

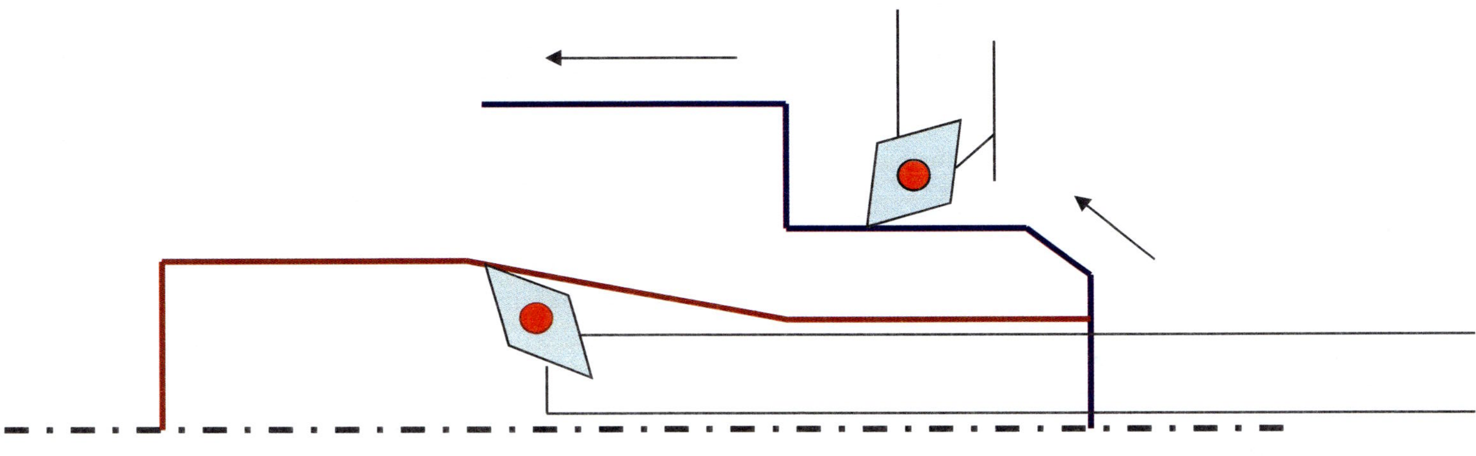

Worksheet 1 CNC Program

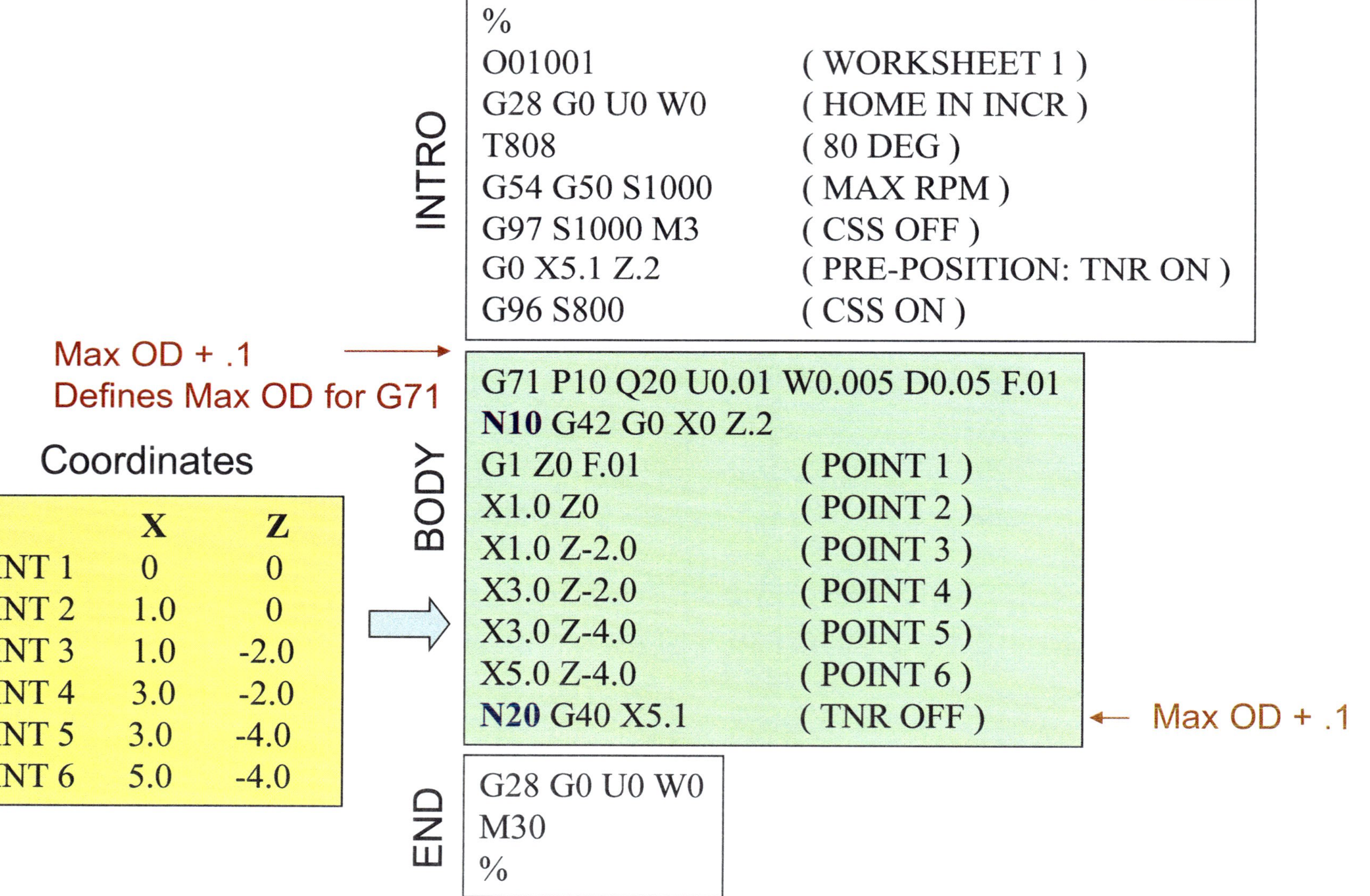

INTRO

```
%
O01001              ( WORKSHEET 1 )
G28 G0 U0 W0        ( HOME IN INCR )
T808                ( 80 DEG )
G54 G50 S1000       ( MAX RPM )
G97 S1000 M3        ( CSS OFF )
G0 X5.1 Z.2         ( PRE-POSITION: TNR ON )
G96 S800            ( CSS ON )
```

Max OD + .1
Defines Max OD for G71

BODY

```
G71 P10 Q20 U0.01 W0.005 D0.05 F.01
N10 G42 G0 X0 Z.2
G1 Z0 F.01          ( POINT 1 )
X1.0 Z0             ( POINT 2 )
X1.0 Z-2.0          ( POINT 3 )
X3.0 Z-2.0          ( POINT 4 )
X3.0 Z-4.0          ( POINT 5 )
X5.0 Z-4.0          ( POINT 6 )
N20 G40 X5.1        ( TNR OFF )
```

Max OD + .1

END

```
G28 G0 U0 W0
M30
%
```

Coordinates

	X	Z
POINT 1	0	0
POINT 2	1.0	0
POINT 3	1.0	-2.0
POINT 4	3.0	-2.0
POINT 5	3.0	-4.0
POINT 6	5.0	-4.0

Mapping Worksheet 2

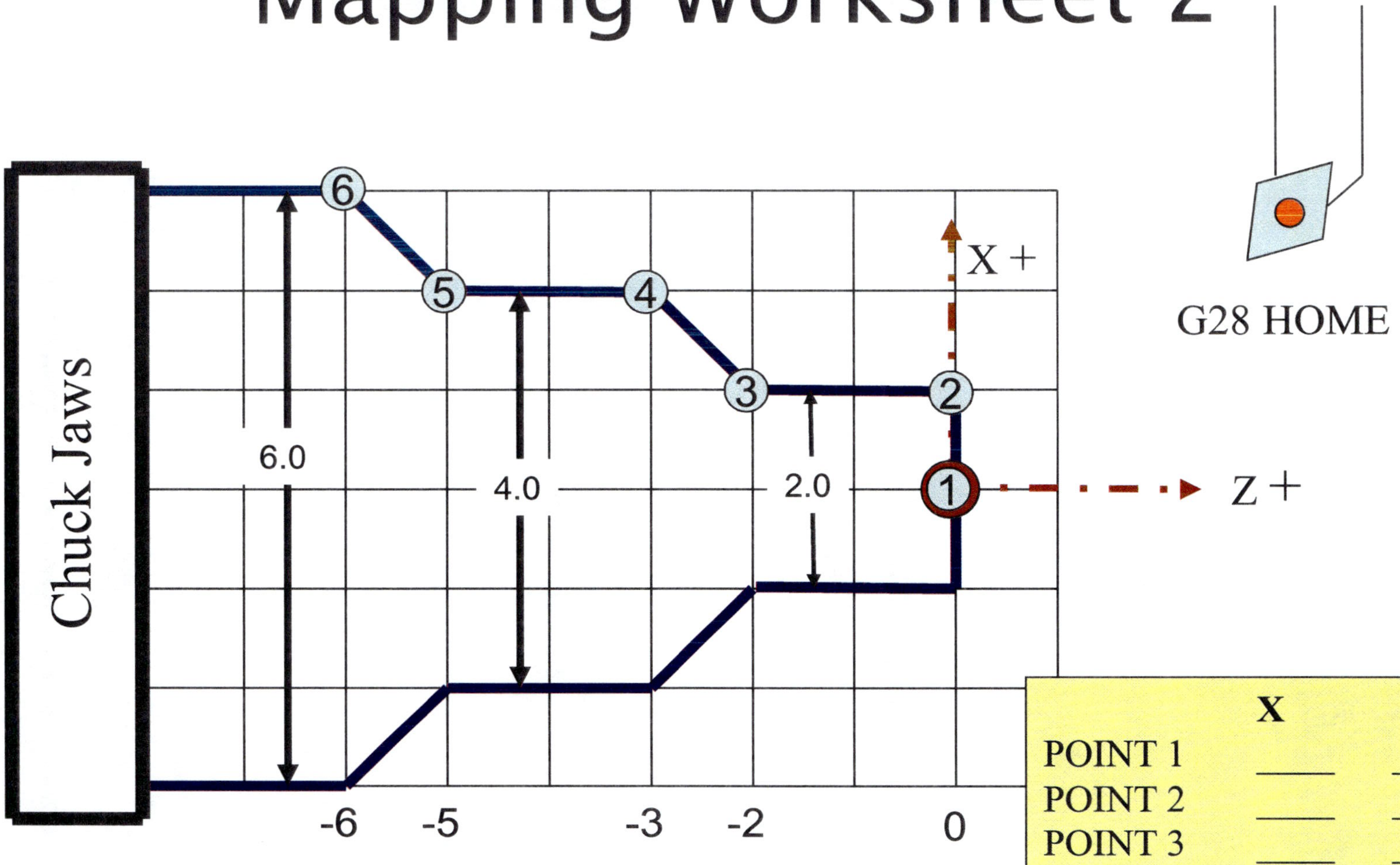

	X	Z
POINT 1	____	____
POINT 2	____	____
POINT 3	____	____
POINT 4	____	____
POINT 5	____	____
POINT 6	____	____

Worksheet 2 Answers

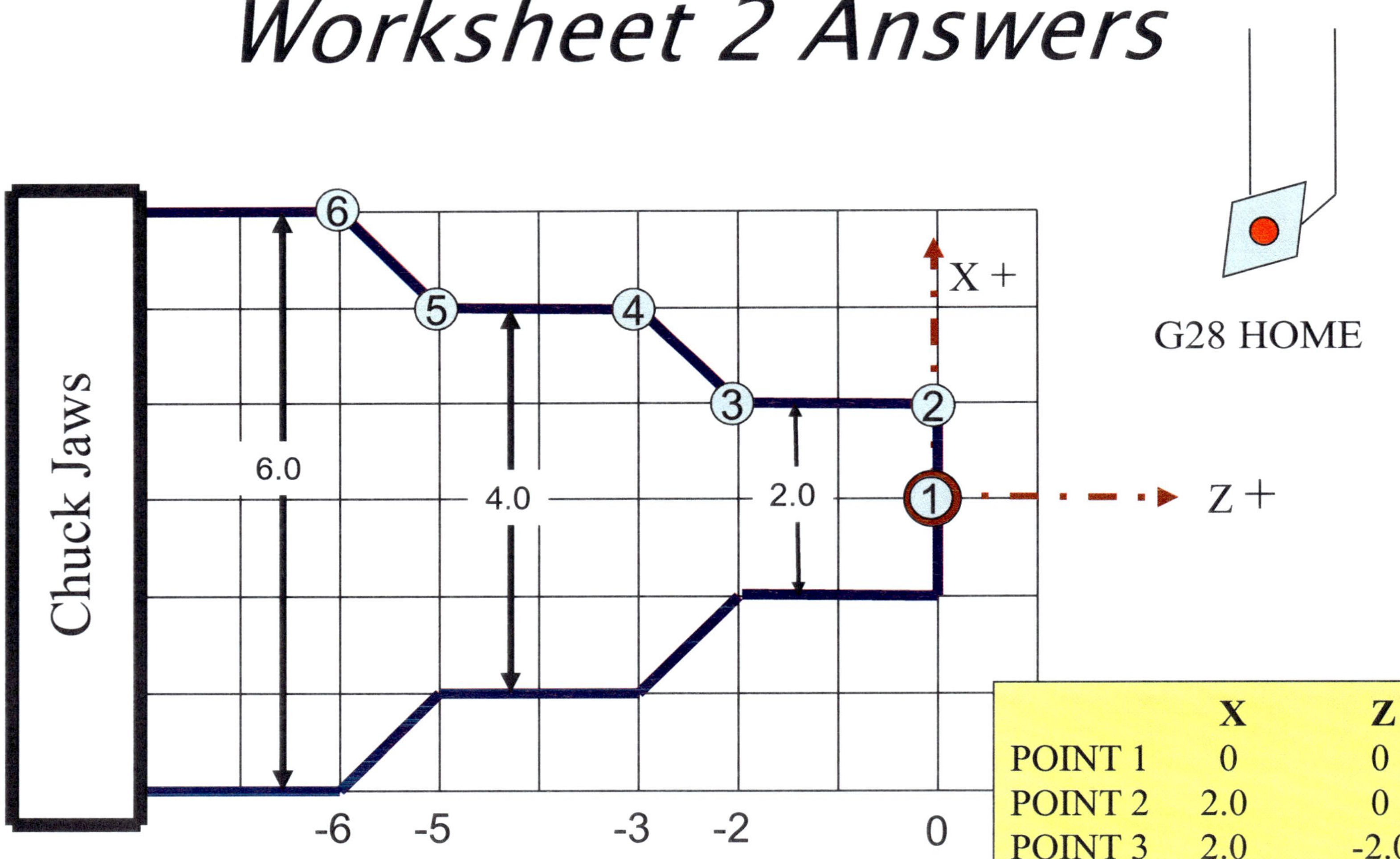

	X	Z
POINT 1	0	0
POINT 2	2.0	0
POINT 3	2.0	-2.0
POINT 4	4.0	-3.0
POINT 5	4.0	-5.0
POINT 6	6.0	-6.0

Worksheet 2 CNC Program

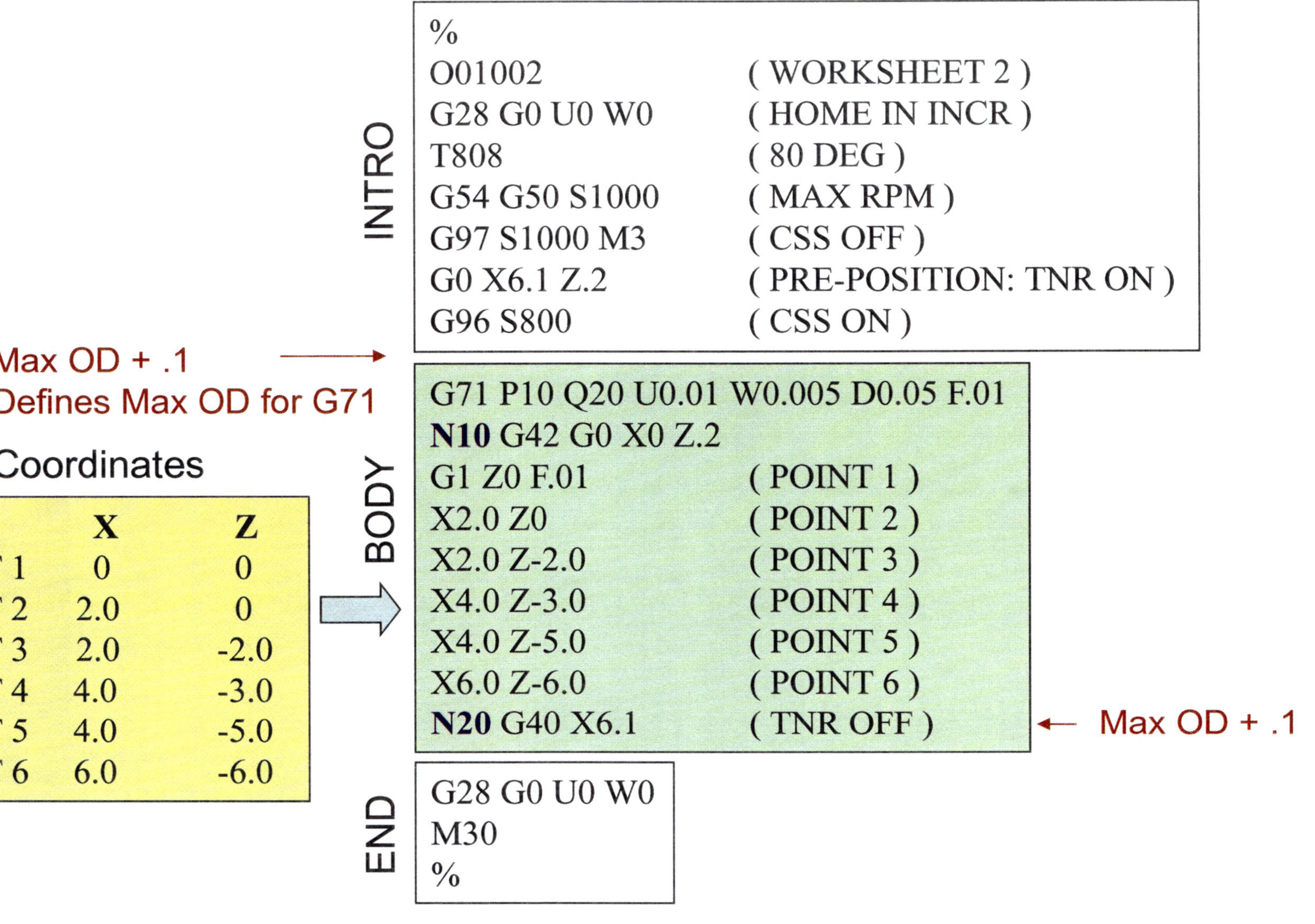

Mapping Worksheet 3

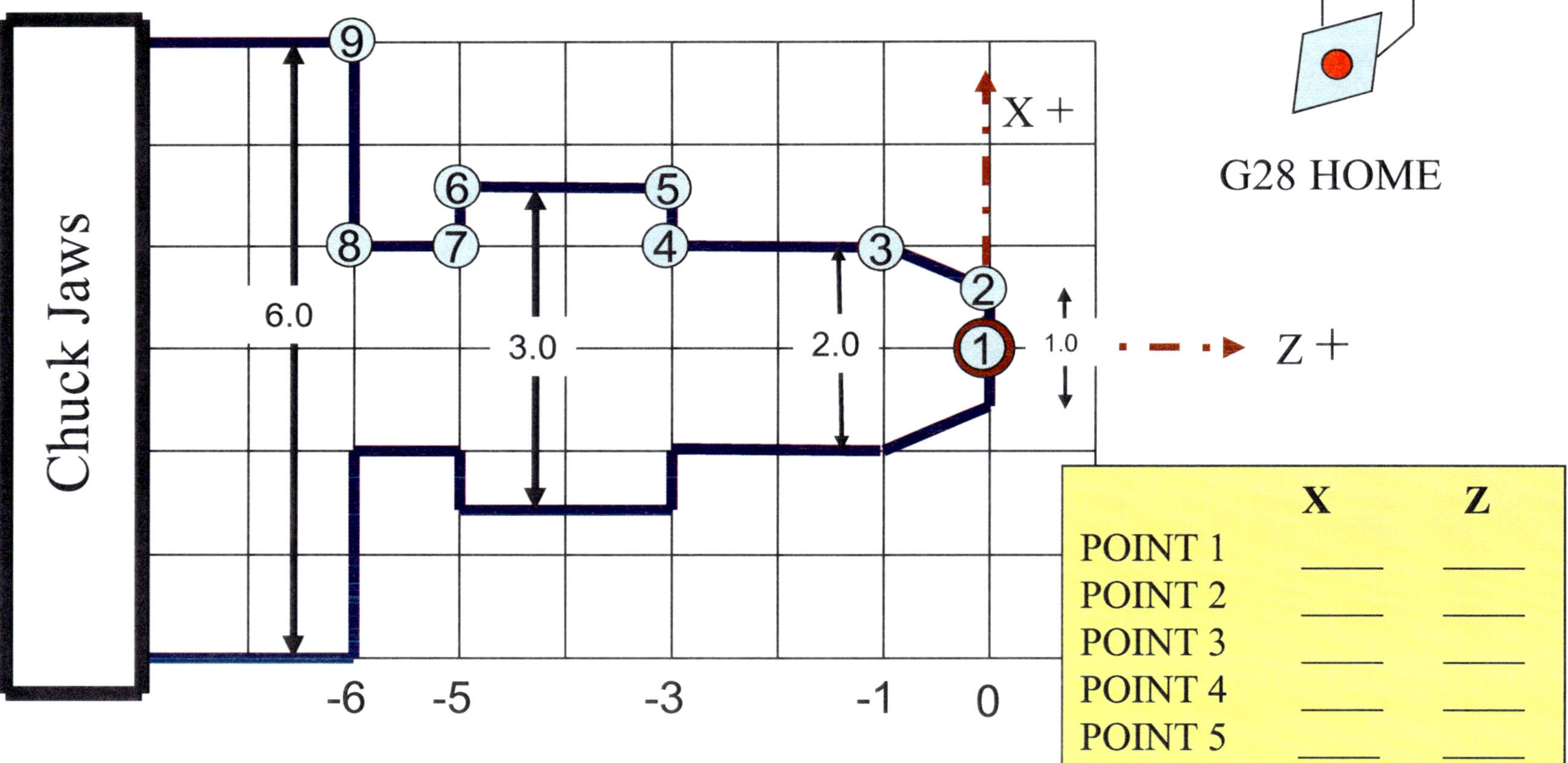

	X	Z
POINT 1	____	____
POINT 2	____	____
POINT 3	____	____
POINT 4	____	____
POINT 5	____	____
POINT 6	____	____
POINT 7	____	____
POINT 8	____	____
POINT 9	____	____

NOTE: In the actual machining process, points 7 and 8 require a secondary grooving operation and grooving tool.

Worksheet 3 Answers

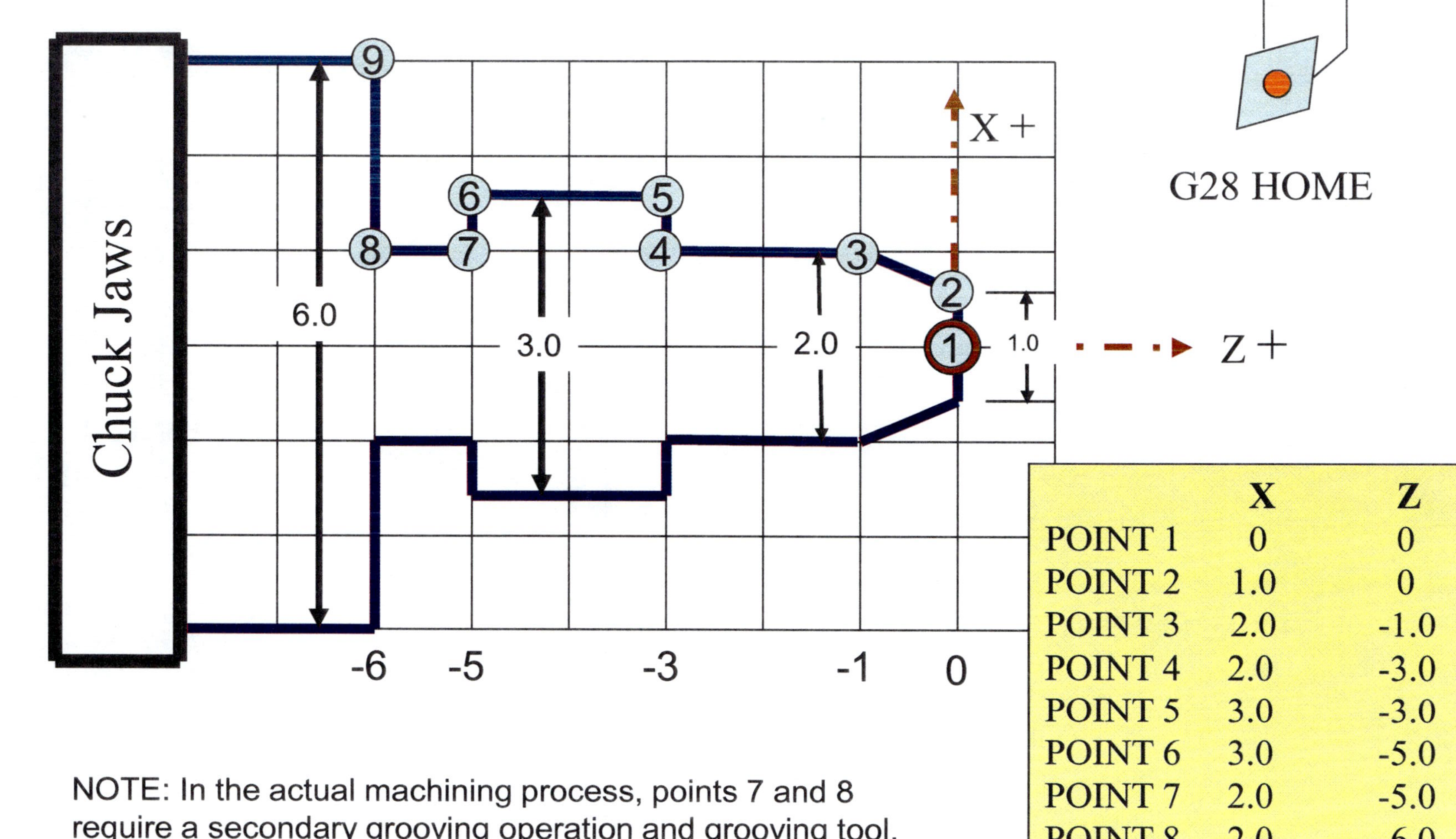

	X	Z
POINT 1	0	0
POINT 2	1.0	0
POINT 3	2.0	-1.0
POINT 4	2.0	-3.0
POINT 5	3.0	-3.0
POINT 6	3.0	-5.0
POINT 7	2.0	-5.0
POINT 8	2.0	-6.0
POINT 9	6.0	-6.0

NOTE: In the actual machining process, points 7 and 8 require a secondary grooving operation and grooving tool.

Worksheet 3 CNC Program

Note:
The following program demonstrates mapping coordinates to the program.
In reality, points 7-8 would not be included in the Rough/Finish cycles. They would be machined later with a groove tool.

Coordinates

	X	Z
POINT 1	0	0
POINT 2	1.0	0
POINT 3	2.0	-1.0
POINT 4	2.0	-3.0
POINT 5	3.0	-3.0
POINT 6	3.0	-5.0
POINT 7	2.0	-5.0
POINT 8	2.0	-6.0
POINT 9	6.0	-6.0

INTRO

```
%
O01003                    ( WORKSHEET 3 )
G28 G0 U0 W0              ( HOME IN INCR )
T808                      ( 80 DEG )
G54 G50 S1000             ( MAX RPM )
G97 S1000 M3              ( CSS OFF )
G0 X6.1 Z.2               ( PRE-POSITION: TNR ON )
G96 S800                  ( CSS ON )
```

BODY

```
G71 P10 Q20 U0.01 W0.005 D0.05 F.01
N10 G42 G0 X0 Z.2
G1 Z0 F.01  ( POINT 1 )
X1.0 Z0                   ( POINT 2 )
X2.0 Z-1.0                ( POINT 3 )
X2.0 Z-3.0                ( POINT 4 )
X3.0 Z-3.0                ( POINT 5 )
X3.0 Z-5.0                ( POINT 6 )
X2.0 Z-5.0                ( POINT 7 )
X2.0 Z-6.0                ( POINT 8 )
X6.0 Z-6.0                ( POINT 9 )
N20 G40 X6.1              ( TNR OFF )
```

END

```
G28 G0 U0 W0
M30
%
```

Chapter 3
Tool Nose Compensation
(TNC)

Objectives

1. The student will know the definition of Tool Nose Compensation (TNC).

2. The student will know the purpose of Tool Nose Compensation (TNC).

3. The student will know the meaning of *uncompensated* and *compensated* as it relates to TNC.

4. The student will know the tool *tip* orientations and when to apply them.

5. The student will know incremental programming works U and W.

Tool Nose Compensation (TNC)

Tool Nose Compensation (TNC) is standard in industry. It allows the tool to be compensated for *wear*, and it allows the programmer to program *print* dimensions, allowing the CNC control to *offset* the tool accordingly. The offset direction is determined by G41 and G42.

TNC Right (G42) compensates the tool to the right side of the profile in the direction of travel—typically used on the Outer Diameter (OD) with motion towards the chuck.

TNC Left (G41) compensates the tool to the left side of the profile in the direction of travel—typically used on the Inner Diameter (ID) with motion towards the chuck.

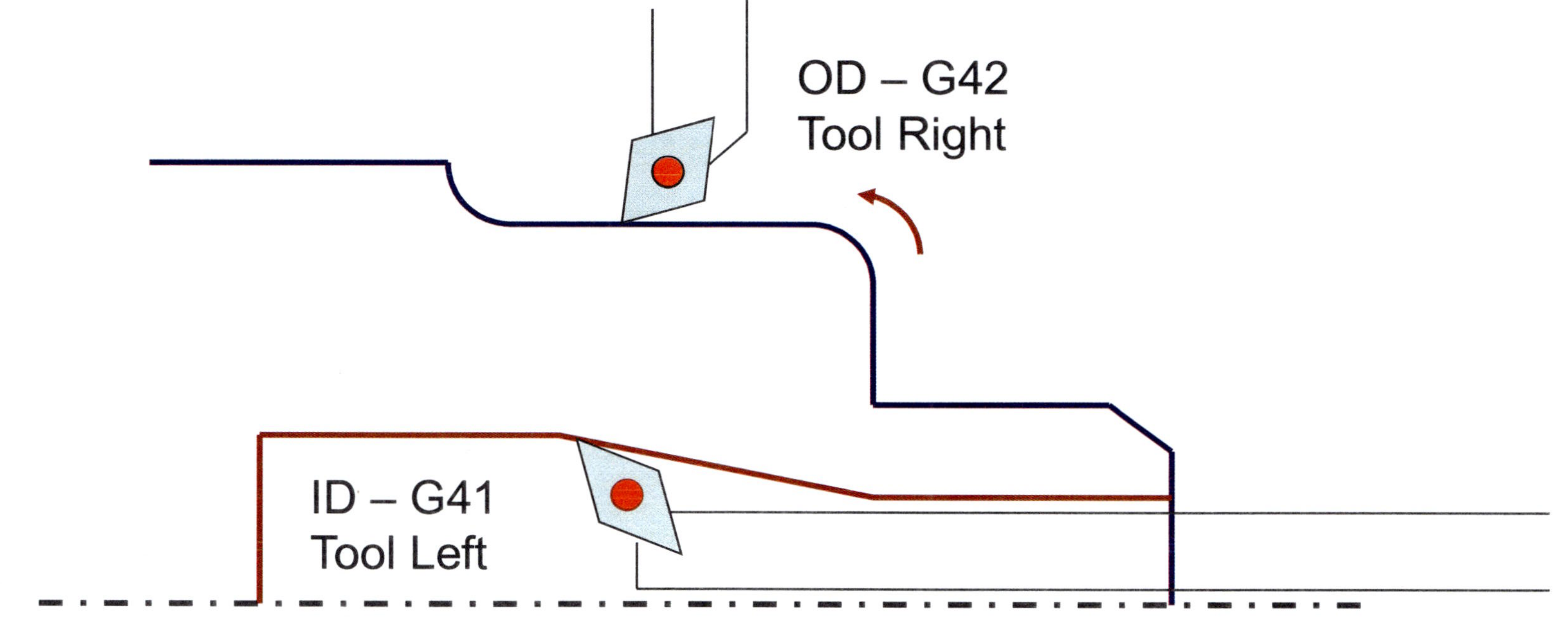

Tool Nose Compensation

Below is a recipe for programming **Tool Nose Compensation**. It's important to know how to turn it on and off properly so the compensation move takes affect away *from* the part.

A fake move is required prior to turning TNR ON to allow room for the tool to move from an uncompensated state to a compensated state.

A fake move is required prior to turning TNR OFF to allow room for the tool to move from a compensated state to an uncompensated state.

A "fake move" is a new cutting move to position the tool into or out of the compensated tool path.

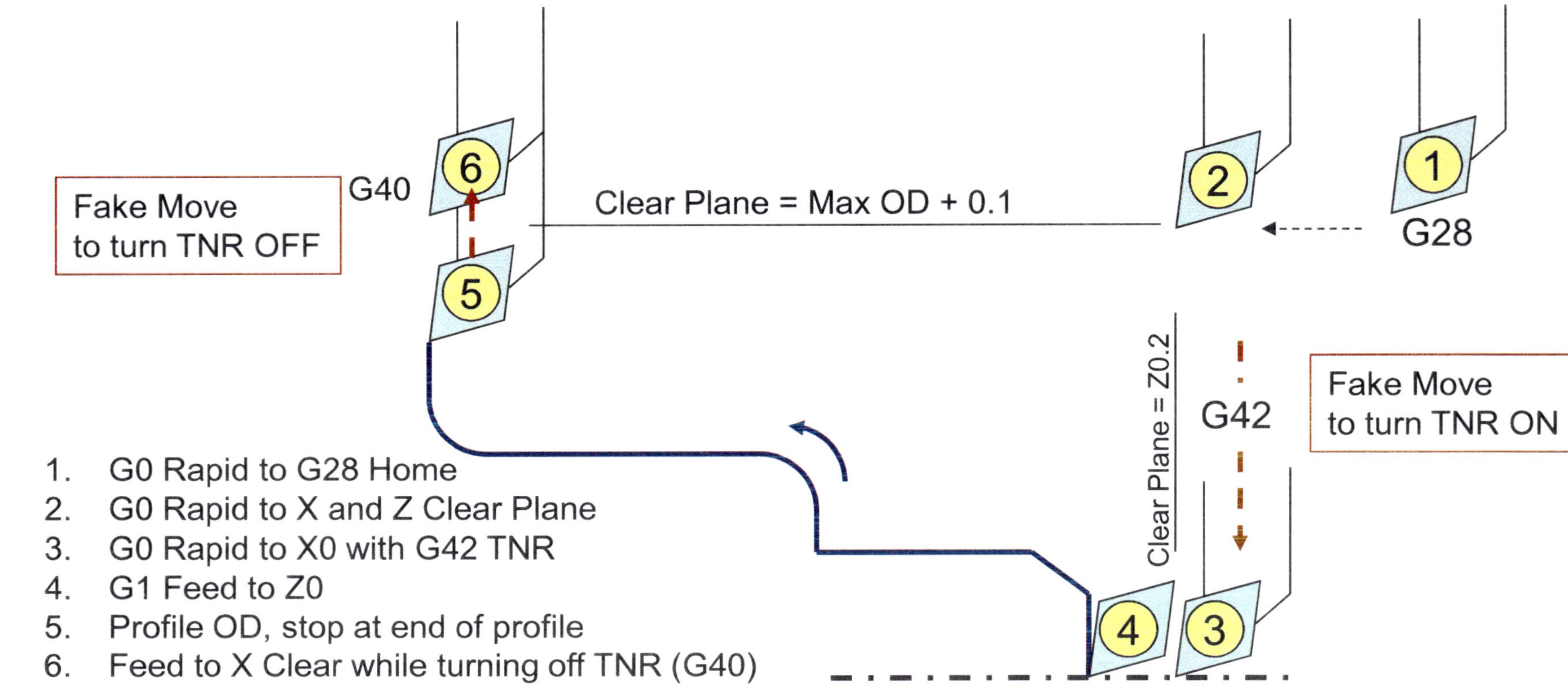

1. G0 Rapid to G28 Home
2. G0 Rapid to X and Z Clear Plane
3. G0 Rapid to X0 with G42 TNR
4. G1 Feed to Z0
5. Profile OD, stop at end of profile
6. Feed to X Clear while turning off TNR (G40)

Tool Tip Orientation

Tool tip orientation must be defined when using TNC. This allows the CNC control to compensate the position of the tool in the proper direction as it machines along a profile.

Tip 3 is the most common tool tip orientation for OD operations.

Tool tip orientation is defined/recorded in the Tool Offset Register in the machine control.

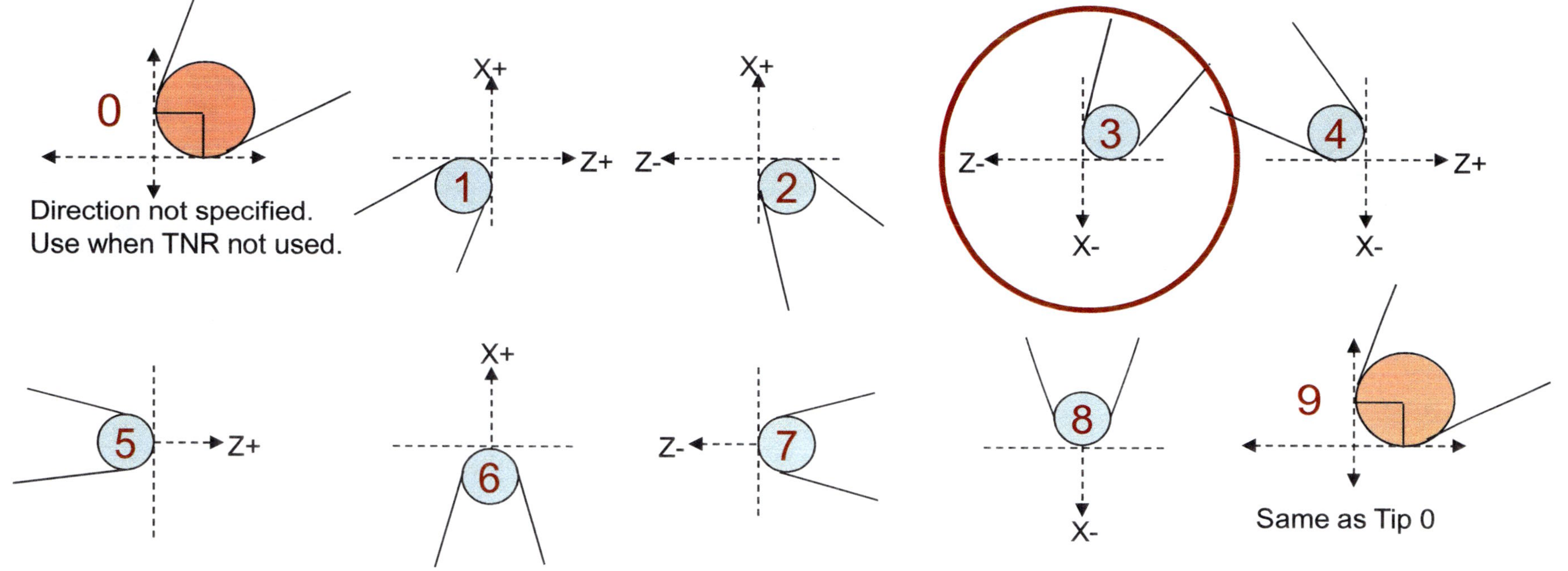

Incremental Programming

CNC Lathes use U and W to program *incremental* motion along their X and Z axis, respectively. G90 and G91 are not used as they are in mill programming. UW programming is sometimes useful for simple motion inside grooves, etc. They're also used with G28 to rapid home.

UW coordinates identify the distance from the previous point.

U: Specifies incremental motion along the X-axis.

W: Specifies incremental motion along the Z-axis.

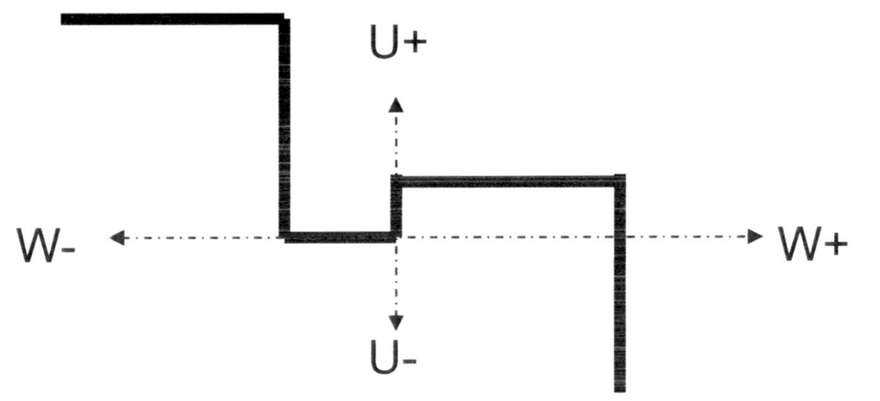

Chapter 4
G & M Codes

Objectives

1. The student will know the common G codes.
2. The student will know the common M codes.
3. The student will know the common CNC codes used in a typical CNC lathe program.
4. The student will know the structure of a typical CNC lathe program.
5. The student will know the definition of Machine Home.

Common G&M Codes

Preparatory Functions

Code	Function
G0	Rapid Positioning
G1	Linear Machining at feed rate
G2	Circular Contouring Clockwise
G3	Circular Contouring CCW
G28	Reference Machine Zero
G40	Cancel TNC
G41	Tool Nose Compensation Left
G42	Tool Nose Compensation Right
G54-G59	Work Offsets
G70	Finish Turn Cycle
G71	Rough Turn Cycle
G76	OD Thread Cycle
G80	Cancel Canned Cycle mode
G81	Drill Canned cycle
G83	Deep Hole Drilling Canned Cycle
G84	RH (Right Hand) Tapping

Miscellaneous Functions

Code	Function
M0	Program Stop
M3	Spindle on CW (forward)
M5	Spindle Stop
M8	Coolant On
M30	End of Program
M97	Call Subroutine
M98	Call Sub Program
M99	Return from Sub

Common CNC Codes

N — **Block Sequence Numbers** are used as a visual aid in the CNC program. N codes are also used to label subroutines and canned cycles.

G — **Preparatory Functions** are used to define a mode, such as linear interpolation (G1), circular contouring (G2 and G3), Drilling (G81), Finish Turn (G70), etc.

XZ — defines **axes coordinates** throughout the program. Coordinates identify a machine's location.

R — defines the **radius** of the arc or circle when circular contouring.

S — **Spindle Speed** specifies how many revolutions per minute the chuck rotates.

F — **Feed rate** specifies the rate of travel expressed in Inches Per Revolution.

T — **Tool** number and offset number. Expressed together, for example T101 (Tool 1, Offset 1).

M — **Miscellaneous Functions** perform operations, such as turn on coolant, end program, etc.

CNC Codes

- CNC codes and values (e.g., G1, M3, S800, X10.0) are programmed in each block.
- Multiple G-codes ***are*** allowed in one block providing they don't conflict with each other. e.g., **G54 G0 G40 G80**
- All other codes can only occur once in a block, including M.
- Some codes are *modal*. Which means they remain in effect until cancelled.
  - They don't have to be repeated in each line. Examples:
    - Linear (G1), Circular (G2 and G3), Canned Cycles (G81), Feed rate (F)
- Integers *vs.* Real Numbers
  - Spindle Speeds, G & M codes, etc., require integer values. e.g., **S1000 M3**
  - X,Y,Z, F, R, and Special Variables must include a decimal point.
    - Failure to include the decimal point causes the decimal point to move four places to the left. e.g., F10 = F.0010

CNC Programs

INTRO

```
%
O01001              ( WORKSHEET 1 )
G28 G0 U0 W0        ( HOME IN INCR )
T808                ( 80 DEG )
G54 G50 S1000       ( MAX RPM )
G97 S1000 M3        ( CSS OFF )
G0 X5.1 Z.2         ( PRE-POSITION: TNR ON )
G96 S800            ( CSS ON )
```

BODY

```
G71 P10 Q20 U0.01 W0.005 D0.05 F.01
N10 G42 G0 X0 Z.2
G1 Z0 F.01          ( POINT 1 )
X1.0 Z0             ( POINT 2 )
X1.0 Z-2.0          ( POINT 3 )
X3.0 Z-2.0          ( POINT 4 )
X3.0 Z-3.0          ( POINT 5 )
X5.0 Z-3.0          ( POINT 6 )
N20 G40 X5.1        ( TNR OFF )
```

END

```
G28 G0 U0 W0
M30
%
```

- CNC programs can be programmed manually or from a CAM system.
  - Manual programs are written in a text editor and should be easy to read and well structured.
  - CAM systems create CNC programs by "***post processing***" programmed toolpaths. They create and format CNC programs according to machine/control conventions. There are thousands of post processors.
- CNC programs are commonly referred to as G&M code programs.
  - They should be created from a simple text editor, such as ***Notepad*** in Windows. When using MS Word, files should be saved as ***Plain Text*** to avoid special characters such as indent, paragraph, etc.
  - CNC programs should be typed in **UPPERCASE**.

CNC Programs

INTRO

```
%
O01001                  ( WORKSHEET 1 )
G28 G0 U0 W0            ( HOME IN INCR )
T808                    ( 80 DEG )
G54 G50 S1000           ( MAX RPM )
G97 S1000 M3            ( CSS OFF )
G0 X5.1 Z.2             ( PRE-POSITION: TNR ON )
G96 S800                ( CSS ON )
```

BODY

```
G71 P10 Q20 U0.01 W0.005 D0.05 F.01
N10 G42 G0 X0 Z.2
G1 F.01 Z0              ( POINT 1 )
X1.0 Z0                 ( POINT 2 )
X1.0 Z-2.0              ( POINT 3 )
X3.0 Z-2.0              ( POINT 4 )
X3.0 Z-3.0              ( POINT 5 )
X5.0 Z-3.0              ( POINT 6 )
N20 G40 X5.1            ( TNR OFF )
```

END

```
G28 G0 U0 W0
M30
%
```

- % is required at the start and end of each program when developed offline.
- O1001 is the program number.
- G28 takes the machine turret to a fixed home position at the start of the program.
- T808 calls up tool 8 and offset 8.
- G50 S1000 sets a maximum speed. Limiting the speed is necessary due to inertia as the spindle speed increases with CSS.
- CSS (Constant Surface Speed) increases the spindle speed as the diameters decrease, and visa versa. G96 turns on CSS. The initial speed is based on the maximum OD. RPM = CS x 3.82 / OD
- G71 is a rough turn cycle.
- The machine turret is sent home at the end of the program.
- M30 ends the program

Block Sequence Numbers

- Each line of code in a CNC program is called a ***block***.
  - Each block can begin with a block sequence number.
  - On most machines they are optional and can be anything you desire.
  - You can increment by 1, 5, 10, or any number you desire to make your program more readable. That is, N10, N20, N30. The spacing between numbers allows you room to insert blocks later if necessary.
  - The maximum number of sequence numbers on most machines is 99999.
  - Many manually written lathe programs do not use N codes.
  - N codes are also used to identify the start and finish block numbers for canned cycles.

HOME

- The turret is commonly sent "*Home*" at the start and end of every program, as well as for Tool Change.
  - "*Home*" is a fixed location on the machine. The absolute coordinate values vary from machine to machine. Therefore, it's common to send a machine "Home" with a G28 command.
- Lathe:
  - G28 G0 U0 W0
  - Commonly used for Tool Change, Start, and End of Program

Main Program Structure

INTRO

```
%
O01030                  ( Impeller Profile)
G28 G0 U0 W0            ( HOME )
T808                    ( 80 DEG )
G54 G50 S1000           ( MAX RPM )
G97 S1000 M3            ( CSS OFF )
G0 X2.5 Z.3             ( PRE-POSITION )
G96 S800                ( CSS ON )
```

BODY

```
G71 P1 Q2 U0.01 W0.005 D0.1 F.01
N1 G42 G0 X0 Z.2
G1 Z0 F0.01
G1 X.430
Z-.177
X.630
X.66 Z-.278
X1.693
Z-.496
G2X2.358 Z-.885 R.776
G1 X2.45
N2 G40 X2.55
G28 G0 U0 W0
G70 P1 Q2
```

END

```
G28 G0 U0 W0
M30
%
```

Impeller

Chapter 5
Feed Rate and Spindle Speed

Objectives

1. The student will know the definitions of Constant Surface Speed and Inches Per Revolution.

2. The student will know the formula for calculating Spindle Speed.

3. The student will know the common tool types used in CNC Turning.

4. The student will know the lathe tool angles.

Feed & Speed

To calculate RPM for turning you must start with recommendations from the tool manufacturer. This data is dependent upon the workpiece material and the material of the cutting tool.

SFPM = SURFACE FEET PER MINUTE, or *Cutting Speed.* SFPM is defined as how many surface feet the circumference of the workpiece travels in one minute. It's primarily driven by the hardness of the workpiece material and the cutting tool material.

IPR = INCHES PER REVOLUTION. IPR is defined as the distance the tool advances with each revolution of the part. Too large of a bite causes the tool to break down. Overheating may be caused by excess RPM (Revolution Per Minute).

SFPM and IPR recommendations come from the tooling manufacturer. The numbers come in a range. Use the low side of the range to be conservative. Use the high side of the range to be more aggressive, or program to the mean and use machine override. It's important, however, to stay within the range of the manufacturer's recommendation to avoid burning up or breaking the tool. Both affect tool life.

$$RPM = \frac{SFPM \times 3.82}{OD}$$

Calculating Low & High Limits

CUTTING SPEED = 400 - 600
OUTSIDE DIAMETER = 1. 5

Low

$$RPM = \frac{400 \text{ X } 3.82}{1.5}$$

$$RPM = \frac{1528}{1.5}$$

RPM = 1018

High

$$RPM = \frac{600 \text{ X } 3.82}{1.5}$$

$$RPM = \frac{2292}{1.5}$$

RPM = 1528

Low & High Limits Worksheet 1

CUTTING SPEED = 550 - 750
OUTSIDE DIAMETER = 1. 125

Low

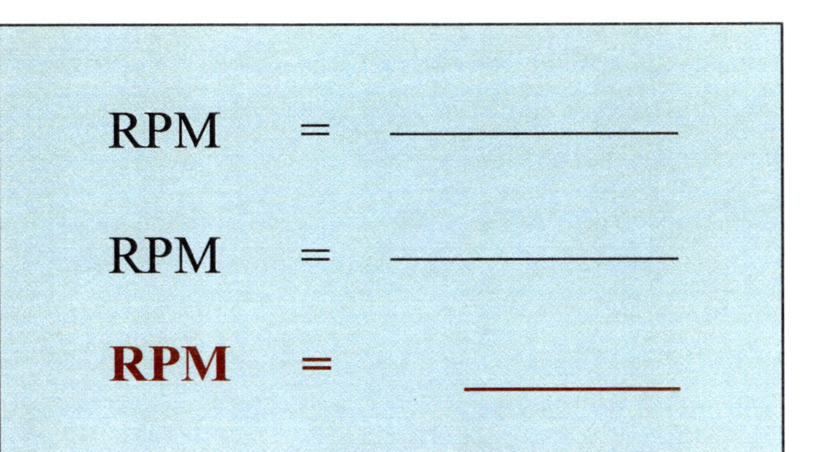

High

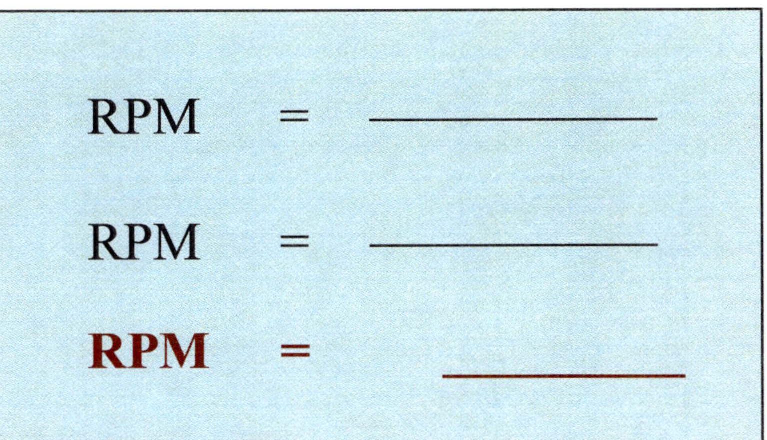

Low & High Limits Worksheet 2

CUTTING SPEED	= 450 - 550
OUTSIDE DIAMETER	= 0. 375

Low

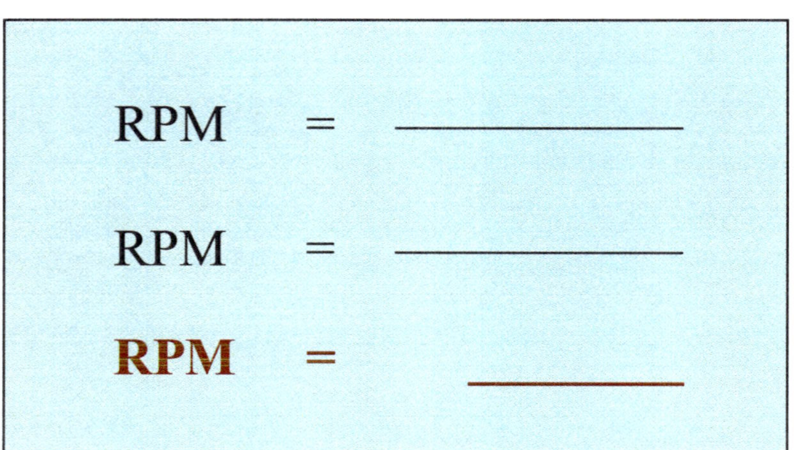

High

RPM = ______

RPM = ______

RPM = ______

Tool Types

A variety of tool shapes exist for CNC turning, largely influenced by the part shape. Tools come in right-hand and left-hand orientations to allow you to turn in both directions. The most common tool is the 80 degree Turn/Face tool. This is used for most operations. However, the side relief angles are too large to reach into small detailed areas, so subsequent tools are typically used.

The cutting and clearance angles of a lathe tool are dependent upon the shape of the part and the quality of the finish required.

Tool Angles

Back Rake affects the ability of the tool to shear the work material. This can be positive or negative. Positive rake reduces the cutting forces and causes less part deflection. However if it's too large of an angle, the tool strength decreases, as well as the capacity to conduct heat. When machining harder parts, the rake angle must be small, or even negative for carbide and diamond inserts. The higher the hardness, the less the rake angle.

In general, the larger the radius of the tool, the stronger the tool. However, larger radii may increase the need for secondary operations to remove the remaining stock.

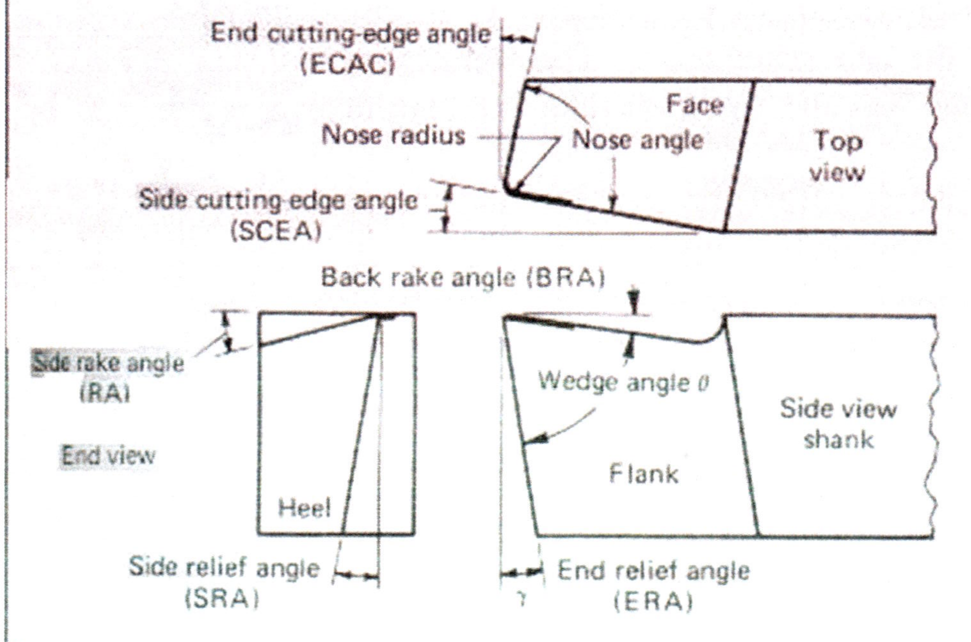

Chapter 6
Circular Interpolation

Objectives

1. The student will know the definition of Circular Interpolation.

2. The student will know how to program Circular Interpolation using R.

Circular Interpolation

- Circular Interpolation allows you to machine arc movements at programmed feed rates.
  - The tool machines from the current position to the programmed position, commanded by a G2 or G3, a desired location (X and/or Z), and a Radius.
    - G3 X2.0 Z-1.5 R0.5
    - G2 and G3 are **modal**. They stay in effect until cancelled by G1 or G0.

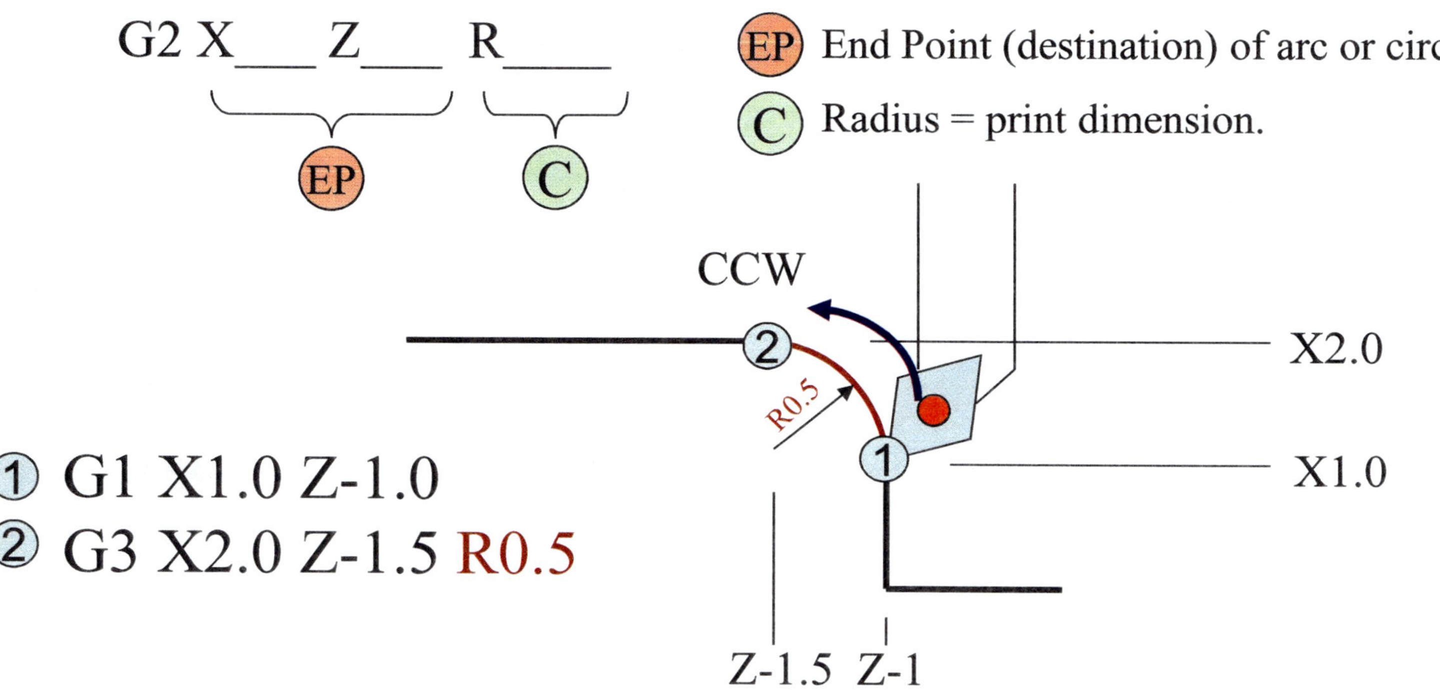

Mapping Print Coordinates with Radii

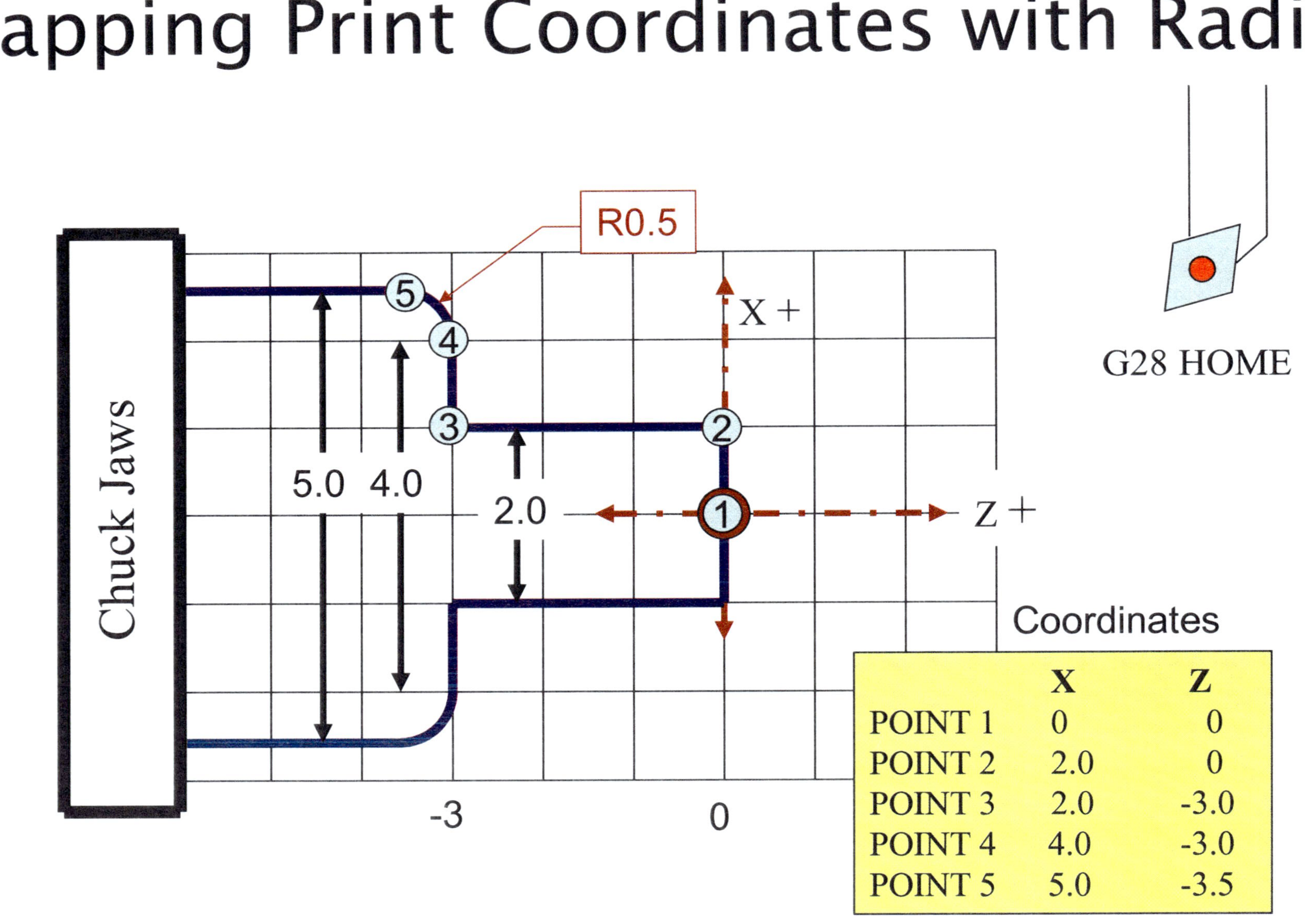

Coordinates

	X	Z
POINT 1	0	0
POINT 2	2.0	0
POINT 3	2.0	-3.0
POINT 4	4.0	-3.0
POINT 5	5.0	-3.5

CNC Program with Circular Interpolation

INTRO

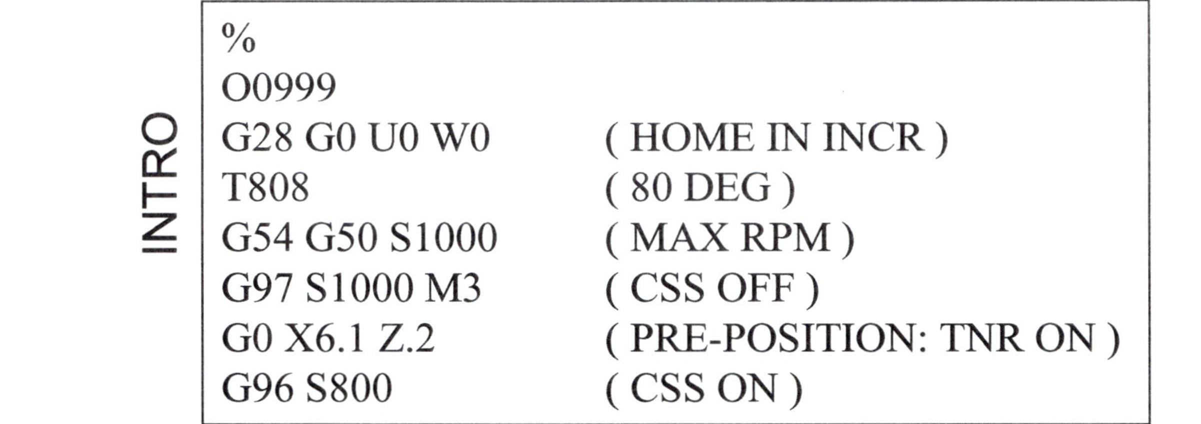

```
%
O0999
G28 G0 U0 W0          ( HOME IN INCR )
T808                  ( 80 DEG )
G54 G50 S1000         ( MAX RPM )
G97 S1000 M3          ( CSS OFF )
G0 X6.1 Z.2           ( PRE-POSITION: TNR ON )
G96 S800              ( CSS ON )
```

Coordinates

	X	Z
POINT 1	0	0
POINT 2	2.0	0
POINT 3	2.0	-3.0
POINT 4	4.0	-3.0
POINT 5	5.0	-3.5

BODY

```
G71 P10 Q20 U0.01 W0.005 D0.05 F.01
N10 G42 G0 X0 Z.2
G1 Z0 F.01            ( POINT 1 )
X2.0 Z0               ( POINT 2 )
X2.0 Z-3.0            ( POINT 3 )
X4.0 Z-3.0            ( POINT 4 )
G3 X5.0 Z-3.5 R0.5    ( POINT 5)
N20 G1 G40 X5.1       ( TNR OFF)
```

END

```
G28 G0 U0 W0
M30
%
```

Chapter 7
Rough and Finish OD Canned Cycles

Objectives

1. The student will know how to remove OD stock using a Rough Canned Cycle.

2. The student will know how to finish turn the OD using the Finish Canned Cycle.

Canned Cycles G70 & G71

- The **G71 Rough Turn** cycle is used to rough barstock down to the finish shape, leaving a specified amount of material on the diameters and faces for finish turning later.
  - The first X motion establishes the diameter of the barstock.
  - G71 contains a *rough* phase and a *finish* phase.
  - Generally the roughing phase consists of repeated passes along the Z-axis. The finish phase follows the programmed path removing the steps remaining from the roughing phase.
- The **G70 Finish Turn** cycle is used to finish turn the profile in a single pass—usually after the G71 cycle.
- The code for both operations is written once, yet executed twice—making for a shorter and more readable program.
- All blocks between P and Q are executed.

Rough Turn G71

G71 P10 Q20 U0.01 W0.005 D0.05 F.01

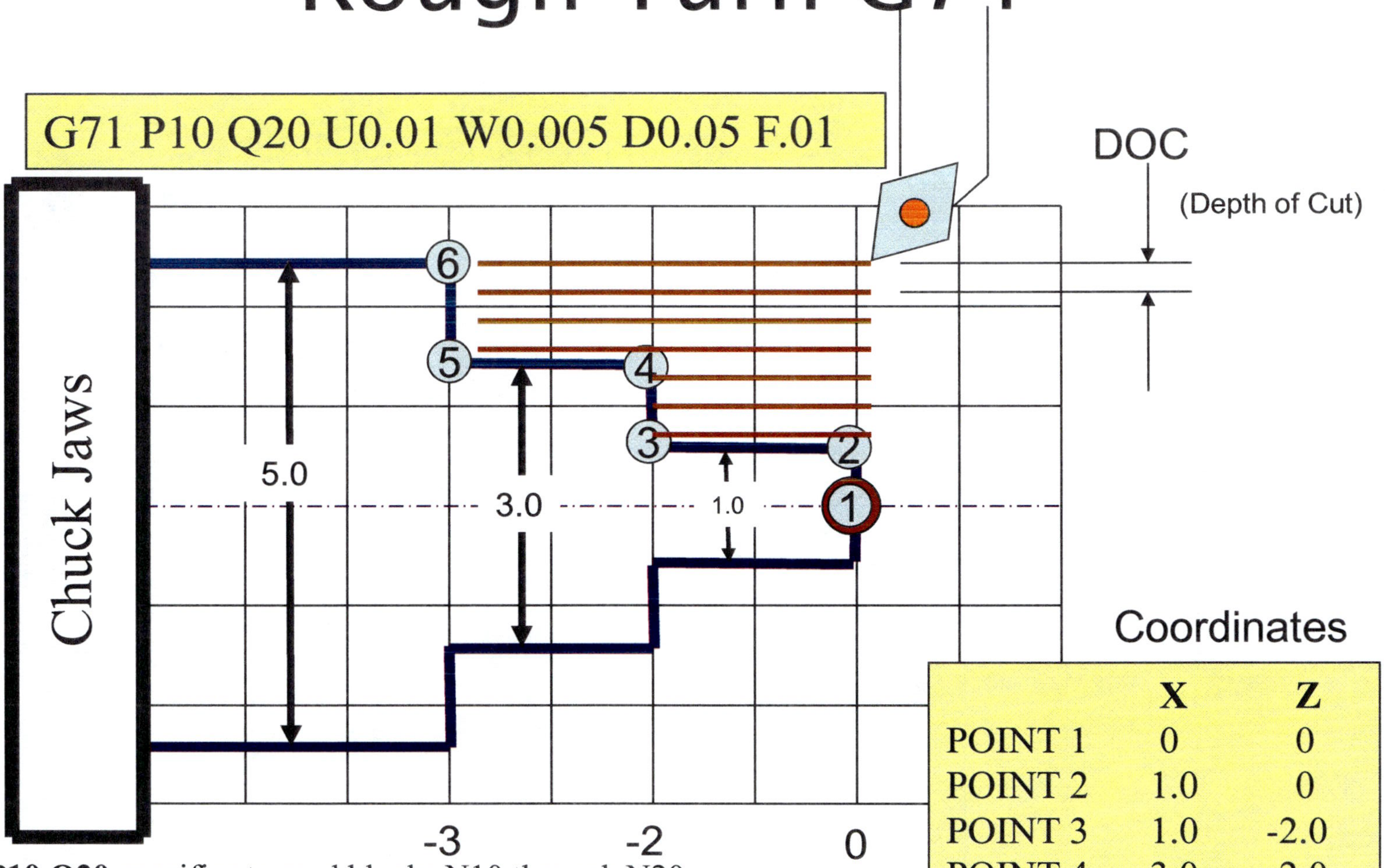

DOC

(Depth of Cut)

Chuck Jaws

5.0

3.0

1.0

-3 -2 0

Coordinates

	X	Z
POINT 1	0	0
POINT 2	1.0	0
POINT 3	1.0	-2.0
POINT 4	3.0	-2.0
POINT 5	3.0	-3.0
POINT 6	5.0	-3.0

P10 Q20 specifies to read blocks N10 through N20.

U0.01 leaves 0.01 stock on all diameters.

W0.005 leaves 0.005 stock on all faces.

D0.05 specifies a 0.05 depth of cut per pass.

F.01 specifies a feed rate of 0.01 while rough turning.

Finish Turn G70

G70 P10 Q20

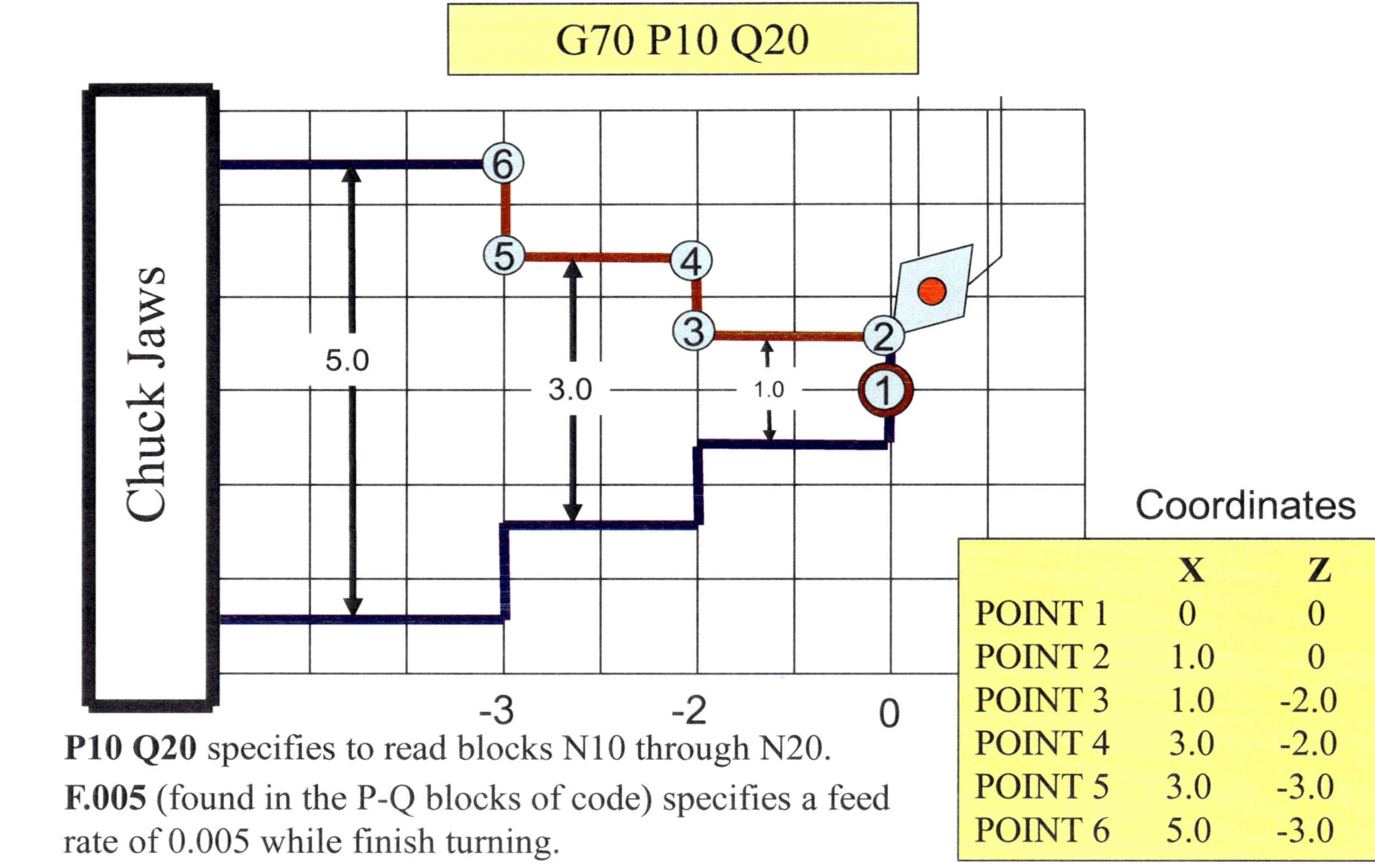

P10 Q20 specifies to read blocks N10 through N20.

F.005 (found in the P-Q blocks of code) specifies a feed rate of 0.005 while finish turning.

Coordinates

	X	Z
POINT 1	0	0
POINT 2	1.0	0
POINT 3	1.0	-2.0
POINT 4	3.0	-2.0
POINT 5	3.0	-3.0
POINT 6	5.0	-3.0

Rough and Finish CNC Program

```
%
O01001              ( WORKSHEET 1 )
G28 G0 U0 W0        ( HOME IN INCR )
T808                ( 80 DEG )
G54 G50 S1000       ( MAX RPM )
G97 S1000 M3        ( CSS OFF )
G0 X5.1 Z.2         ( PRE-POSITION: TNR ON )
G96 S800            ( CSS ON )
```

Max OD + .1
Defines Max OD for G71

```
G71 P10 Q20 U0.01 W0.005 D0.05 F.01
N10 G42 G0 X0 Z.2
G1 Z0 F.005         ( POINT 1 )
X1.0 Z0             ( POINT 2 )
X1.0 Z-2.0          ( POINT 3 )
X3.0 Z-2.0          ( POINT 4 )
X3.0 Z-3.0          ( POINT 5 )
X5.0 Z-3.0          ( POINT 6 )
N20 G40 X5.1        ( TNR OFF )
```

Max OD + .1

```
G28 G0 U0 W0
T1010
G70 P10 Q20
```

Coordinates

	X	Z
POINT 1	0	0
POINT 2	1.0	0
POINT 3	1.0	-2.0
POINT 4	3.0	-2.0
POINT 5	3.0	-3.0
POINT 6	5.0	-3.0

Chapter 8
OD Thread Canned Cycle

Objectives

1. The student will define thread terminology.

2. The student will know how to calculate depth of thread.

3. The student will know how to define feed rate for thread cycles.

4. The student will know how to program threads using the Thread Canned Cycle.

G76 External Threading Cycle

- The **G76 Threading cycle** is used to produce straight threads on the outside diameter (OD) of the part in multiple passes.
- TNC is not used when threading.
- G99 (Feed Per Revolution) is programmed with feed rate to synchronize the feed rate with spindle speed.

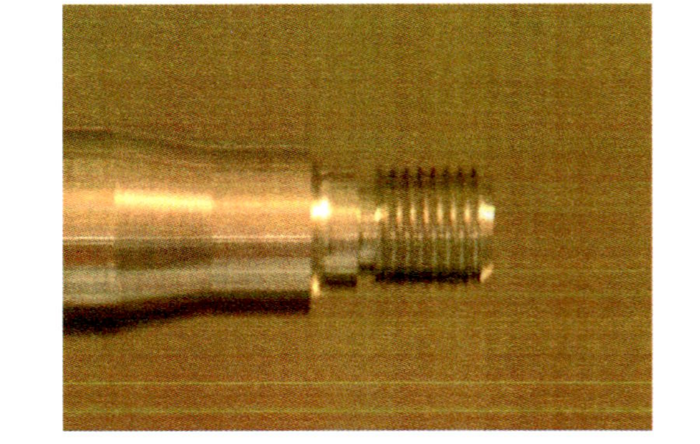

G76 X0.8466 Z-2.0 K0.0767 D0.008 F0.125

X = Minor Diameter = Major Thread Diameter – (2 x K)

Z = Length of thread along the Z-axis.

K = Height of Thread: 0.6134 or 0.6495 / Number of Threads

D = Depth of cut for first pass. (Must be smaller than K)

F = 1 / Threads Per Inch = Pitch

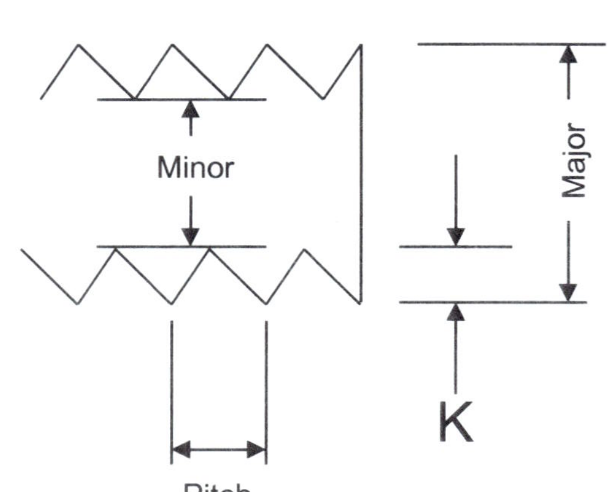

G76 Worksheet 1–8 Thread

Develop a canned cycle to produce an external thread with a **1-8 thread**.

G76 X0.8466 Z-2.0 K0.0767 D0.008 F0.125

Major Diameter = 1.0
Threads Per Inch = 8

$$\mathbf{K} = \frac{0.6134}{\text{Threads Per Inch}} = \frac{0.6134}{8} = 0.0767$$

$$\textbf{Feed rate} = \frac{1}{\text{Threads Per Inch}} = \frac{1}{8} = 0.125$$

$$\begin{aligned} \mathbf{X}\ \text{Minor} &= \text{Major Diameter} - (2 \times K) \\ &= 1.0 - (2 \times 0.0767) \\ &= 1.0 - 0.1534 \\ &= 0.8466 \end{aligned}$$

NOTE: 0.6495 can be used in place of 0.6134

G76 Worksheet ¾–10 Thread

Develop a canned cycle to produce an external thread with a **3/4-10** thread.

G76 X_____ Z-2.0 K_____ D0.008 F______

Major Diameter = _____

Threads Per Inch = _____

K = $\dfrac{0.6134}{\text{Threads Per Inch}}$ = ________ = ____________

Feed rate = $\dfrac{1}{\text{Threads Per Inch}}$ = $\dfrac{1}{\quad}$ = ____________

X Minor = Major Diameter - (2 x K)

=

=

=

NOTE: 0.6495 can be used in place of 0.6134

G76 Worksheet ½–13 Thread

Develop a canned cycle to produce an external thread with a **1/2-13 thread**.

G76 X______ Z-2.0 K_____ D0.008 F______

Major Diameter = ______

Threads Per Inch = ______

$$\mathbf{K} = \frac{0.6134}{\text{Threads Per Inch}} = \frac{\quad}{\quad} = \_\_\_\_\_\_$$

$$\textbf{Feed rate} = \frac{1}{\text{Threads Per Inch}} = \frac{1}{\quad} = \_\_\_\_\_\_$$

X Minor = Major Diameter - (2 x K)

=

=

=

NOTE: 0.6495 can be used in place of 0.6134

G76 Sample Code

1” – 8 Thread

G28 G0 U0 W0

T505 (Thread Tool)

G54 G50 S1000

G97 S1000 M3

G0 X1.0 Z1.0 (Position the tool to start)

Z0.1 M24 (Z start point, don’t cut chamfer at end of thread)

G76 X.8466 Z-2.0 K.0767 D.008 F0.125

(K = 0.6134 / 8)

(X = 1 - (K X 2)

G0 X1.1   (Clear Plane)

G28 G0 U0 W0

Chapter 9
Haas CNC Lathe
Setup and Operation

Objectives

1. The student will know how to power on and off the HAAS CNC lathe.

2. The student will know the primary methods of inputting a program.

3. The student will know how to simulate (graph) programs on the control.

4. The student will know how to touch off tools in X and Z.

Turning On The Machine

- Power On
- RESET
- Power Up/Restart
  - ***Note***: You're required to close (cycle) the doors prior to Power Up.

CNC Program Input

- There are three primary methods of inputting a program:

1. Type it into the control using MDI
   - List Program > (Program Number) O'n' > Type your program
2. Floppy, Memory Stick, Network:
   - Word Processor
   - CAM (post processed files)
   - Downloaded from control's hard drive or memory
3. RS232 (DNC)

1. Manual Data Input (MDI)

- LIST PROGRAM
  - Type program number O1234 Press WRITE/ENTER
    - Loads existing program if this number exists, or...
    - Creates new program
  - Edit
    - Create and edit program (Enter a Line – Press WRITE/ENTER)
    - Note: everything is stored when the machine is powered off.
- To save your MDI program:
  - Press HOME to position the cursor to the top of the screen.
  - Press O'n' and ALTER. (e.g., O1111 then ALTER
    - This clears the screen and puts program into memory or your hard drive.
  - Press LIST program to locate the new file and select it.

2. Load from Floppy/Memory/Network

- Word Processors are used to manually type in CNC programs offline. However, the processed CNC program must be saved in in a ***Plain Text*** format:
  - Notepad or WordPad
    - ***Note 1***: you must save your file as Text Only or Plain Text. Saving in a Plain Text format will remove unnecessary formatting.
    - ***Note 2***: Your programs should be free of indentations, footers, headers, etc. Be especially careful not to add any special formatting.
- To Load Programs on machines **without** USB/Network option:
  - LIST PROG
  - Type in desired program to load (e.g., O1234)
  - F3
- To list contents from floppy on machines without network option:
  - LIST PROG
  - F4 (reads floppy)
  - O8999 (dummy program)
  - Overwrite? Y
  - When complete, EDIT lists contents of floppy

2. Load from Floppy/Memory/Network (*cont.*)

- To list contents from a device on machines equipped **with** USB/Network option:
  - LIST PROG
  - Select appropriate device from the list on the left (Hard disk, USB, etc.).
  - F4 reads and displays contents on screen
  - Be sure input focus is on the right side of the screen, then using your up and down arrow keys, select the appropriate program to load.
  - WRITE/ENTER
  - Left Arrow
  - Cursor up to memory or hard drive
  - F2 copies the file
  - ***Note***: Before ejecting your USB stick, LIST PROG and press ORIGIN. Failure to do this may cause harm to the USB stick.

- **3. RS-232:** Programs can be uploaded/downloaded to the machine using a null-modem cable connected from your PC to the machine. You'll need a communications package and you must match the settings in the HAAS control with the settings in the communications package.

Graphing (2D Simulation)

- The HAAS controls have the ability to graph (simulate) your programs. It's recommended to graph all programs prior to machining.
  - Select Program
  - Press SETTING GRAPH twice
  - Press CYCLE START

- To Zoom:
  - F2 (Activates Zoom functions)
  - PAGE DOWN (Zoom In)
  - PAGE UP (Zoom Out)
  - Arrow keys (position zoom box over part)
  - WRITE/ENTER to accept zoom position

Lathe Tool Touch Off

- Load part in chuck
  - Open chuck jaws by depressing left foot pedal
  - Load stock
  - Close jaws by depressing left foot pedal
- ***Note***: The right foot pedal brings in the tailstock. Should you bring it in by mistake, press the **TS** button located near the JOG area of the control panel until it's fully retracted.

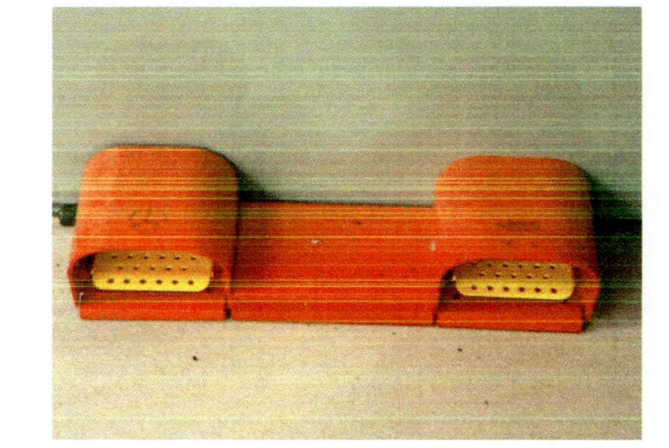

- **Touch off 80 degree Turn/Face tool**
  - Turret Fwd to Tool 8 or...
  - MDI: T808 CYCLE START
  - MDI: T808 Turret Fwd

- **Touch off Z**
  - Turn on spindle rotation
  - Using Hand Jog, jog to the *face* of the part
    - Use .1 for quick movement; .001 for slow movement
  - Face the part to 80% cleanup.
  - Press OFFSET to Tool Offset Page
    - Tool 8 (Position cursor to Z Field)
    - Press Z FACE MEASURE
      - Watch the field to be sure it takes effect.
    - Jog away from part in X and Z

- **Touch off X**
  - Turn on spindle rotation
  - Using Hand Jog, jog to the *diameter* of the part
    - Use .1 for quick movement; .001 for slow movement
  - Skim (turn) a small area of the part near the end.
    - When complete, jog back in Z—not X
    - STOP the spindle or RESET
    - Measure the diameter with micrometer or calipers
  - Press OFFSET to Tool Offset Page
    - Tool 8 (Position cursor to X Field)
    - Press X DIAMETER MEASURE
      - Input diameter. (CANCEL deletes one character at a time).
      - WRITE/ENTER (watch field to be sure it changes)
    - Jog away from part in X and Z
  - [HOME G28]
  - Repeat for all tools necessary

APPENDIX A
Lathe CNC Programming
Level 1 Projects

Lathe CNC Project 1

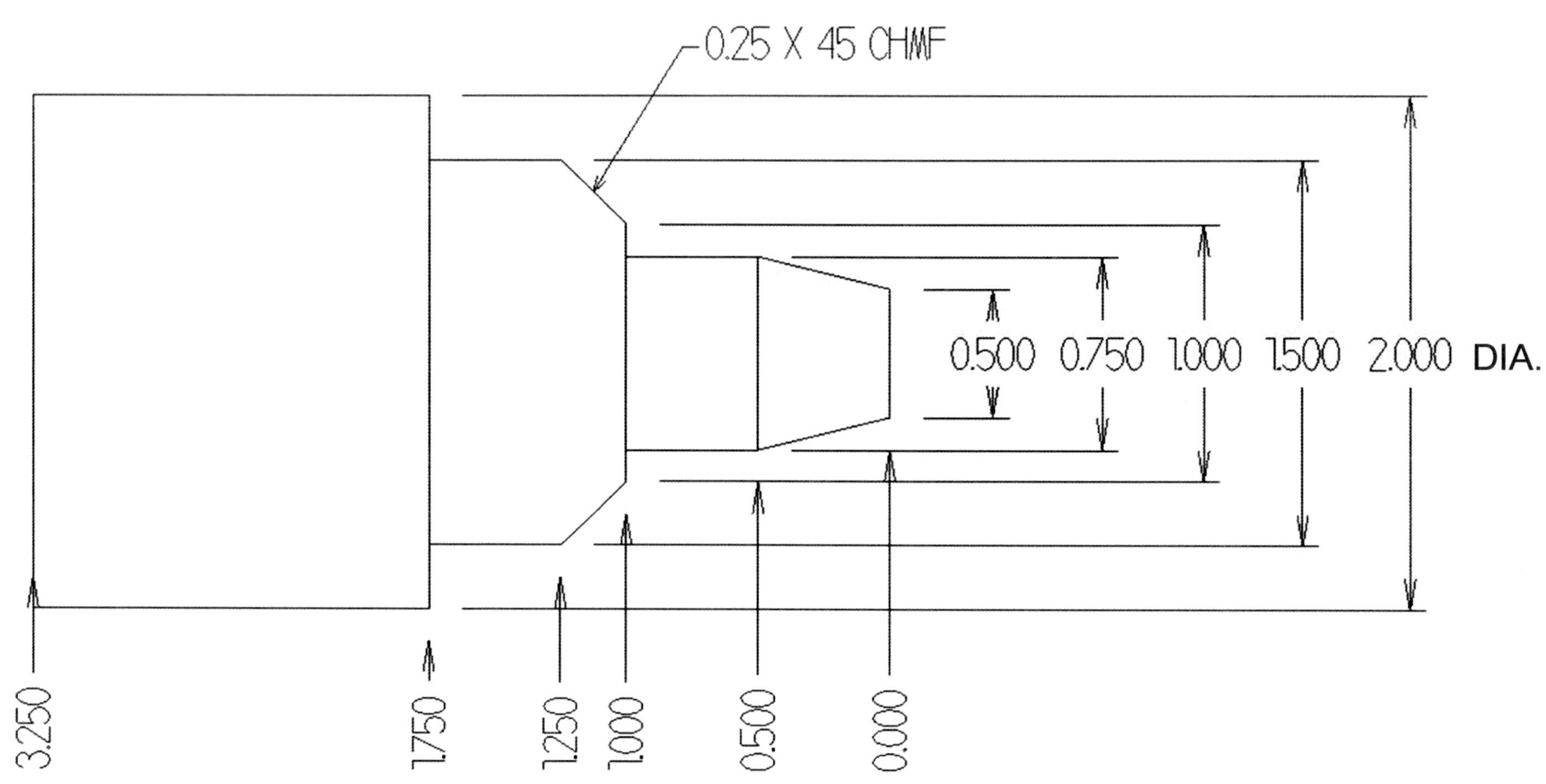

Lathe CNC Project 2

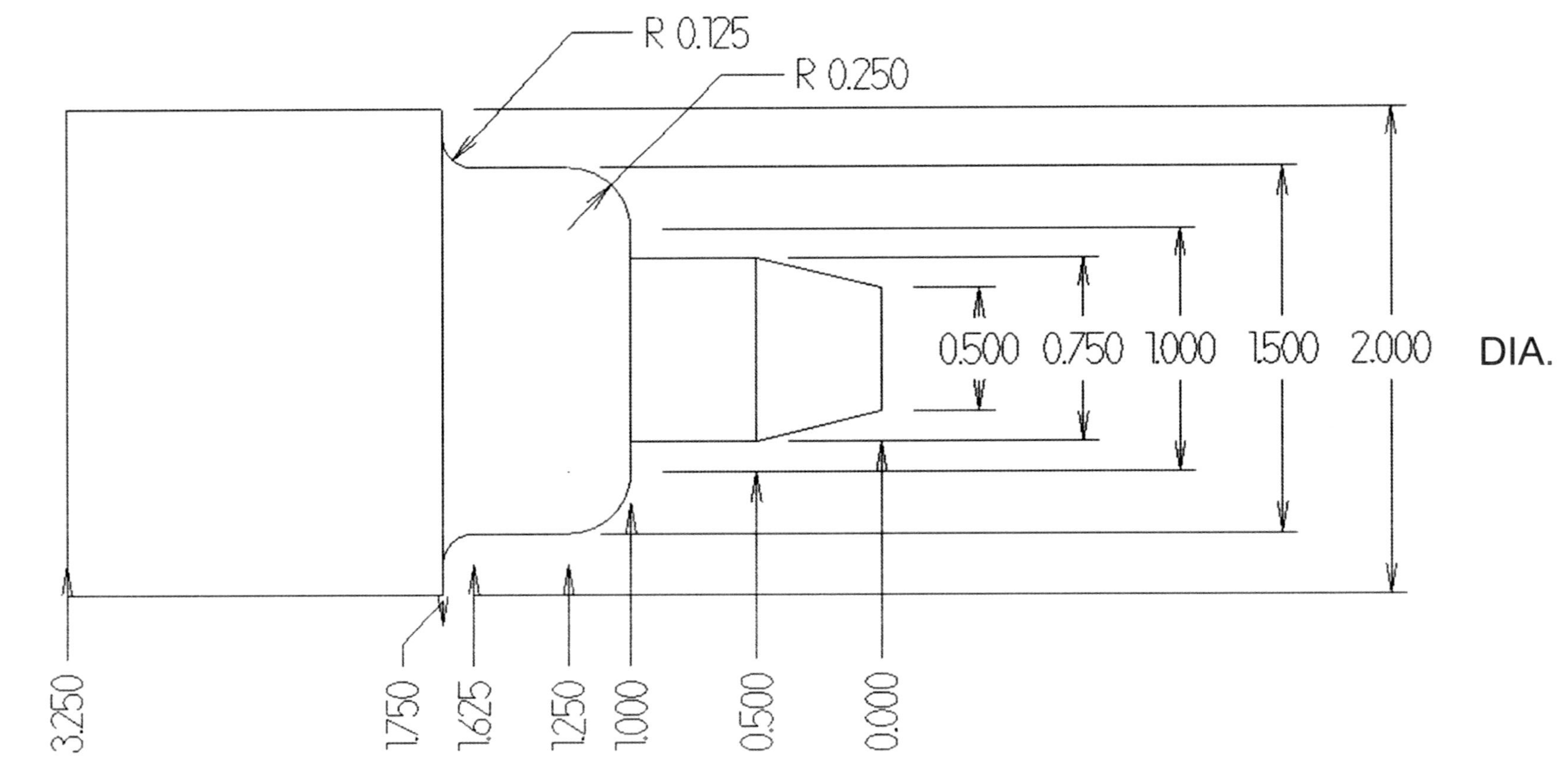

Lathe CNC Project 3

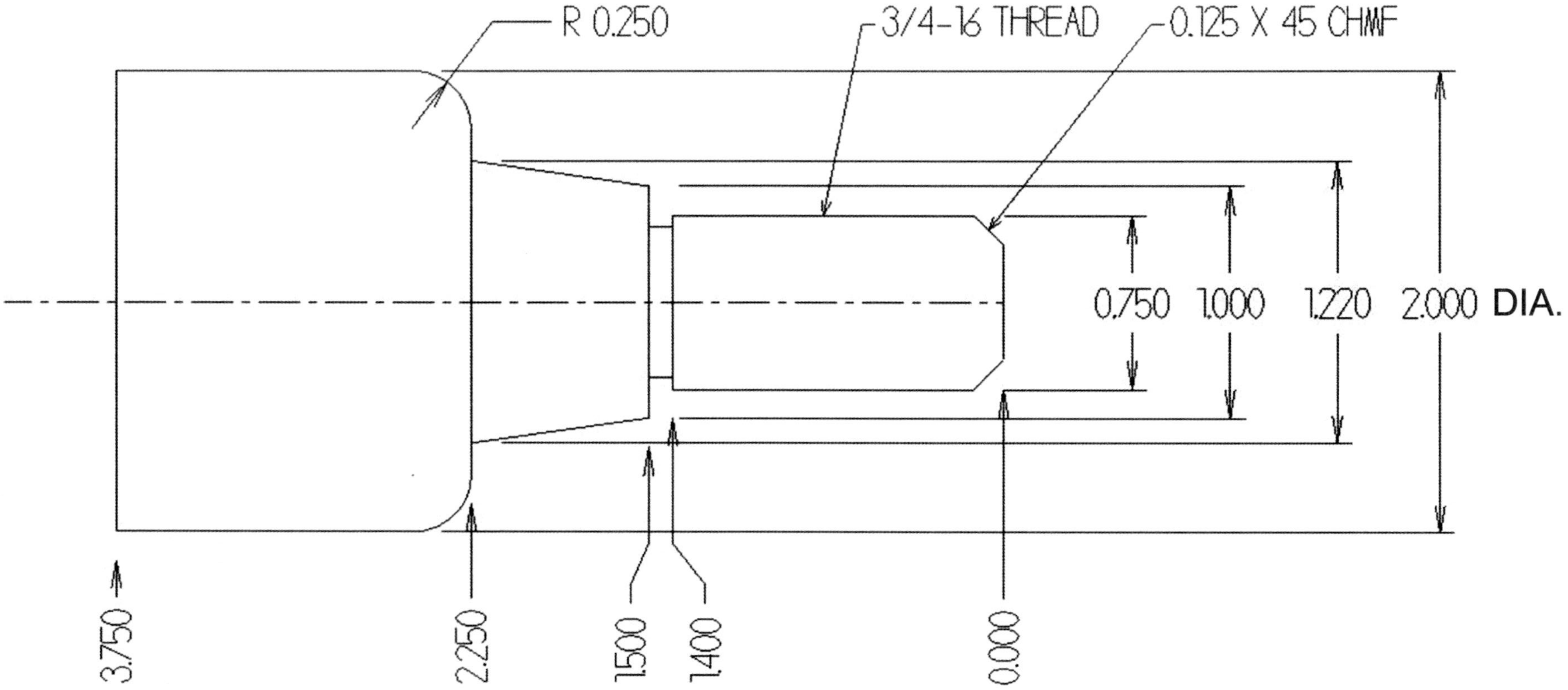

R 0.250
3/4-16 THREAD
0.125 X 45 CHMF
0.750 1.000 1.220 2.000 DIA.
3.750
2.250
1.500
1.400
0.000

Lathe CNC Project 4

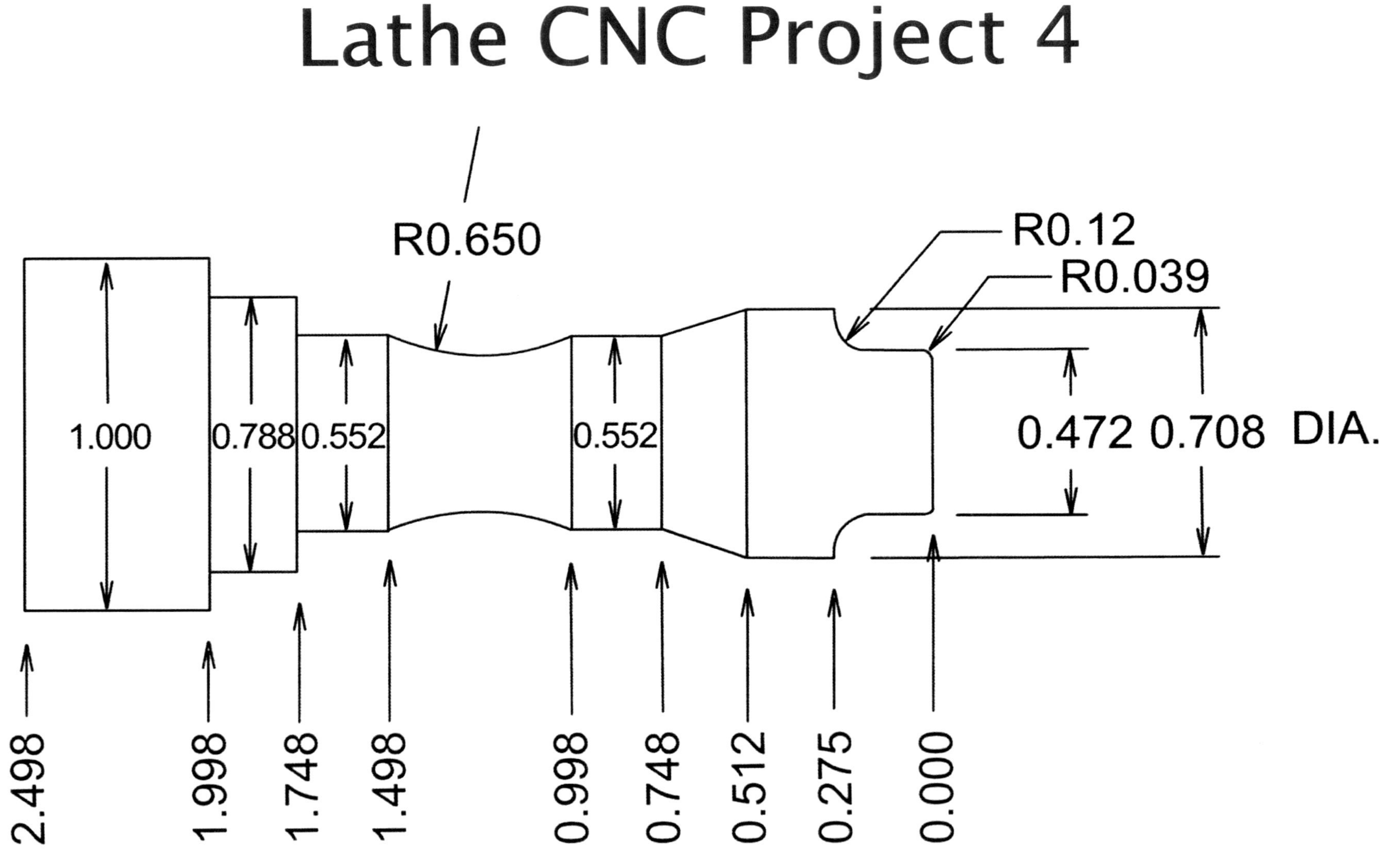

Answers and Programs

Answers and Programs

Worksheets

Worksheet 1 (pg. 13)

```
%
O01001                          (WORKSHEET 1)
G28 G0 U0 W0                    ( HOME IN INCR )
T0808              ( 80 DEG )
G54 G50 S1000   ( MAX RPM )
G97 S1000 M3                    ( CSS OFF )
G0 X5.1 Z.2 ( PREPOSITION: TNR ON )
G96 S800       ( CSS ON )
G71 P10 Q20 U0.01 W0.005 D0.05 F.01

N10 G42 G0 X0 Z.2
G1 Z0 F.01      ( POINT 1 )
X1.0 Z0         ( POINT 2 )
X1.0 Z-2.0      ( POINT 3 )
X3.0 Z-2.0      ( POINT 4 )
X3.0 Z-4.0      ( POINT 5 )
X5.0 Z-4.0      ( POINT 6 )
N20 G40 X5.1 ( TNR OFF )

G28 G0 U0 W0
M30
%
```

Worksheet 2 (pg. 17)

```
%
O01002                          ( WORKSHEET 2 )
G28 G0 U0 W0                    ( HOME IN INCR )
T0808              ( 80 DEG )
G54 G50 S1000   ( MAX RPM )
G97 S1000 M3                    ( CSS OFF )
G0 X6.1 Z.2 ( PREPOSITION: TNR ON )
G96 S800       ( CSS ON )
G71 P10 Q20 U0.01 W0.005 D0.05 F.01

N10 G42 G0 X0 Z.2
G1 Z0 F.01      ( POINT 1 )
X2.0 Z0         ( POINT 2 )
X2.0 Z-2.0      ( POINT 3 )
X4.0 Z-3.0      ( POINT 4 )
X4.0 Z-5.0      ( POINT 5 )
X6.0 Z-6.0      ( POINT 6 )
N20 G40 X6.1 ( TNR OFF )

G28 G0 U0 W0
M30
%
```

Answers and Programs

Worksheets

Worksheet 3 (pg. 20)

```
%
O01003                        ( WORKSHEET 3 )
G28 G0 U0 W0                  ( HOME IN INCR )
T0808               ( 80 DEG )
G54 G50 S1000       ( MAX RPM )
G97 S1000 M3                  ( CSS OFF )
G0 X6.1 Z.2 ( PREPOSITION: TNR ON )
G96 S800       ( CSS ON )
G71 P10 Q20 U0.01 W0.005 D0.05 F.01


N10 G42 G0 X0 Z.2
G1 Z0 F.01     ( POINT 1 )
X1.0 Z0        ( POINT 2 )
X2.0 Z-1.0     ( POINT 3 )
X2.0 Z-3.0     ( POINT 4 )
X3.0 Z-3.0     ( POINT 5 )
X3.0 Z-5.0     ( POINT 6 )
X2.0 Z-5.0     ( POINT 7 )
X2.0 Z-6.0     ( POINT 8 )
X6.0 Z-6.0     ( POINT 9 )
N20 G40 X6.1 ( TNR OFF )

G28 G0 U0 W0
M30
%
```

Answers and Programs

Impeller

Impeller Program (pg. 38)

```
%
O01030                    ( Impeller Profile)
G28 G0 U0 W0              ( HOME )
T808                      ( 80 DEG )
G54 G50 S1000             ( MAX RPM )
G97 S1000 M3              ( CSS OFF )
G0 X2.5 Z.3    ( PRE-POSITION )
G96 S800       ( CSS ON )
G71 P1 Q2 U0.01 W0.005 D0.1 F.01
N1 G42 G0 X0 Z.2
G1 Z0 F0.01
G1 X.430
Z-.177
X.630
X.66 Z-.278
X1.693
Z-.496
G2X2.358 Z-.885 R.776
G1 X2.45
N2 G40 X2.55
G28 G0 U0 W0
G70 P1 Q2
G28 G0 U0 W0
M30
%
```

Answers and Programs

Worksheets

Low and High Worksheet (pg. 43)

Low RPM = 1867
High RPM = 2546

G76 Thread Worksheet ¾ - 10 (pg. 62)

K = 0.0613 (rounded up)
F = 0.1
x = 0.6273

Low and High Worksheet (pg. 44)

Low RPM = 4584
High RPM = 5602

G76 Thread Worksheet ½ - 13 (pg. 63)

K = 0.0944 (rounded up)
F = 0.077 (rounded up from 0.0769)
x = 0.4056

Answers and Programs

Projects

Project 1 (pg. 77)

```
%
O2001                         ( PROJECT 1)
G28 G0 U0 W0                  ( HOME IN INCR )
T0808                ( 80 DEG TURN/FACE )
G54 G50 S1000  ( MAX RPM )
G97 S1000 M3                  ( CSS OFF )
G0 X2.1 Z.2    ( PREPOSITION: TNR ON )
G96 S800       ( CSS ON )
G71 P10 Q20 U0.01 W0.005 D0.05 F.01

N10 G42 G0 X-.062 Z.2 (START BELOW CL)
G1 Z0 F.01
X0.5
X0.75 Z-0.5    (1st taper   )
Z-1.0
X1.0
X1.5 Z-1.25    (.25 CHAMFER  )
Z-1.75
X2.0
N20 G40 X2.1 ( TNR OFF    )

G28 G0 U0 W0
M30
%
```

Project 2 (pg. 78)

```
%
O2002                         ( PROJECT 2)
G28 G0 U0 W0                  ( HOME IN INCR )
T0808                ( 80 DEG TURN FACE )
G54 G50 S1000  ( MAX RPM )
G97 S1000 M3                  ( CSS OFF )
G0 X2.1 Z.2    ( PREPOSITION: TNR ON )
G96 S800       ( CSS ON )
G71 P10 Q20 U0.01 W0.005 D0.05 F.01

N10 G42 G0 X-.062 Z.2 (START BELOW CL)
G1 Z0 F.008
X0.5
X0.75 Z-0.5    (1st taper   )
Z-1.0
X1.0
G3 X1.5 Z-1.25 R0.25  (.25  RADIUS )
G1 Z-1.625
G2 X1.75 Z-1.75 R0.125 (0.125 RADIUS )
G1 X2.0
N20 G40 X2.1 ( TNR OFF    )

G28 G0 U0 W0
M30
%
```

Answers and Programs

Projects

Project 3 (pg. 79)

```
%
O2004                    ( PROJECT 3)
G28 G0 U0 W0             ( HOME IN INCR )
T0808            ( 80 DEG TURN FACE )
G54 G50 S1000   ( MAX RPM )
G97 S1000 M3             ( CSS OFF )
G0 X2.1 Z.2     ( PREPOSITION: TNR ON )
G96 S800        ( CSS ON )

(### ROUGH TURN OD WITH TNR ### )
G71 P10 Q20 U0.01 W0.005 D0.05 F.01
N10 G42 G0 X-.062 Z.2
G1 Z0 F.008
X0.5
X0.75 Z-0.125   ( 1st CHAMFER   )
Z-1.5
X1.0
X1.220 Z-2.25   ( Taper )
X1.5          ( start of radius )
G3 X2.0 Z-2.5 R0.25
N20 G1 G40 X2.1          ( TNR OFF     )
G0 Z0.2         ( MOVE TO CLEAR )

(### FINISH TURN OD ### )
G70 P10 Q20
G28 G0 U0 W0

( ### OD GROOVE WITHOUT TNR ### )
T202 (Groove Tool)
G97 S1000 M3
G0 X2.1 Z-1.5 (RAPID TO START GROOVE)
G1 X.55 F.010 (CUT THE GROOVE)
G1 X1.05     (RETRACT TO CLEAR)
G28 G0 U0 W0

(### THREAD OD WITH NO TNR ### )
T303
G97 S1000 M3
G0 X.750 Z1.0
Z0.2 M24        ( DON'T CUT CHAMFER )
G76 X0.673 Z-1.490 K0.0383 D0.008 F0.0625
G40 X2.1
G28 G0 U0 W0
M30
%
```

Answers and Programs

Projects

Project 4 (pg. 80)

```
%
O203              (Chess Piece: KING   )
( MATERIAL:           ALUMINUM            )
( STOCK SIZE:         1" BARSTOCK x 3" LONG)
( EXTEND FROM JAWS:   2"                  )
( MAX LENGTH OF CUT:  1.998               )
( TOOL:              55 DEGREE DIAMOND   )
( SFM:               1000                )

G28 G0 U0 W0
T909
G54 G50 S3800        (MAX RPM)
G97 S3000 M3         (CSS OFF)
G0 X1.1 Z.2          (PREPOSITION: TNR ON)
G96 S3000            (CSS ON)
G71 P10 Q20 U.01 W.005 D.035 F.01
N10 G42 G0 X-.062 Z.2
G1 Z0 F.008
X.394
G3 X.472 Z-.039 R.039
G1 Z-.175
G2 X.708 Z-.275 R.12
G1 Z-.512
X.552 Z-.748         ( Taper )
Z-.998
G2 Z-1.498 R.65      (LARGE RADIUS)
G1 Z-1.748
G1 X0.788
Z-1.998
G1 X1.0
N20 G1 G40 X1.1      (TURN TNR OFF)
G0 Z.2

G70 P10 Q20
G28 G0 U0 W0
M30
%
```